A SPECIAL ISSUE OF
COGNITIVE NEUROPSYCHOLOGY

Pure Alexia
(Letter-by-letter Reading)

edited by

Max Coltheart

Macquarie University, Sydney, NSW, Australia

Copyright © 1998 by Psychology Press Ltd, a member of the Taylor & Francis group
All rights reserved. No part of this publication may be reproduced, stored in a retrieval system, or transmitted in any form or by any means, electronic, mechanical, photocopying, recording or otherwise, without permission in writing from the publisher.

Psychology Press Ltd, Publishers
27 Church Road
Hove
East Sussex
BN3 2FA
UK

British Library Cataloguing-in-Publication Data

A catalogue record for this book is available from the British Library

ISBN: 0-86377-999-9

Typeset by Quorum Technical Services Ltd, Cheltenham
Printed and bound in the United Kingdom by Henry Ling Ltd, Dorchester

Contents

1   Max Coltheart, *Seven Questions About Pure Alexia (Letter-by-letter Reading)*

7   Marlene Behrmann, David C. Plaut, and James Nelson, *A Literature Review and New Data Supporting an Interactive Account of Letter-by-letter Reading*

53   Martin Arguin, Daniel Bub, and Jeffrey Bowers, *Extent and Limits of Covert Lexical Activation in Letter-by-letter Reading*

93   Marie Montant, Tatjana A. Nazir, and Michel Poncet, *Pure Alexia and the Viewing Position Effect in Printed Words*

141   Eleanor M. Saffran and H. Branch Coslett, *Implicit vs. Letter-by-letter Reading in Pure Alexia: A Tale of Two Systems*

167   Doriana Chialant and Alfonso Caramazza, *Perceptual and Lexical Factors in a Case of Letter-by-letter Reading*

203   Michele Miozzo and Alfonso Caramazza, *Varieties of Pure Alexia: The Case of Failure to Access Graphemic Representations*

239   Subject Index

This book is also a double special issue of the journal *Cognitive Neuropsychology* which forms Issues 1 and 2 of Volume 15 (1998).

Is the Disorder Homogeneous?

This is actually two questions, both to do with the issue of whether LBL reading is a syndrome in the sense of a set of symptoms which invariably co-occur and which have a single common cause.

The first of these two questions is: Are there any abnormalities other than the letter-by-letter reading behaviour which will inevitably be seen in every LBL reader? CC answer in the affirmative because they, like Farah and Wallace (1991) and others, attribute LBL reading behaviour to a low-level visual impairment not specific to the domain of reading. From this view it follows that sufficiently careful assessment of low-level visual processing with stimulus materials unrelated to reading will reveal that LBL reading behaviour is always accompanied by another symptom—a nonspecific impairment of low-level visual processing. MPN agree with this view, suggesting that "the fundamental impairment in LBL reading . . . is a general perceptual deficit that degrades the quality of visual input" and adding that while it is "beyond the scope of this paper . . . whether the poor letter processing derives from an even more fundamental perceptual problem", this is a view that "we (amongst others) have argued elsewhere (Behrmann et al., 1997; Sekuler & Behrmann, 1996)."

SC's answer to this question is somewhat different. For them, the left hemisphere damage leading to LBL reading does not degrade the letter-processing performed in that hemisphere—the damage abolishes such processing. On their view, the letter processing involved in LBL reading is carried out by intact right-hemisphere structures. This does not commit them to any view about how LBL patients will perform in tests of low-level visual processing that do not involve letter-processing, and hence their view does not imply that the symptom of LBL reading will always be accompanied by the symptom of visual perceptual impairment on some nonreading task.

The second question concerning homogeneity is this: Is the cause of the letter-by-letter reading behaviour the same in all patients? BPN, ABB, MNP, and SC all take the view that this is so (even though they differ as to what this cause is). CC are not so sure: "Although the various accounts of LBL reading briefly reviewed here are often discussed as alternative explanations of the disorder (e.g. Behrmann & Shallice, 1995), it is entirely possible that they each constitute plausible explanations for different forms of a heterogeneous reading disorder".

Is the Relevant Impairment Specific to the Reading System or Is it a More General Visual impairment?

This question is central to the study of LBL reading; I have just discussed it in a different context, and indicated that CC and BPN incline to the view that the abnormal reading is a manifestation of a more general visual perceptual impairment that is not specific to the reading system, whereas SC are prepared to attribute the disorder to an impairment or disconnection of a reading-specific letter-processing system in the left hemisphere.

What Is the Nature and the Locus of the LBL Reader's Impairment, Within Some Functional Model of Reading?

Both BPN and ABB suggest that the nature of this impairment is abnormally weak activation within the normal reading system (rather than qualitatively different input to it originating from the right hemisphere, which is the SC view). BPN discuss various possible loci of such weakness of activation within the context of the Interactive Activation Model (IAM) of visual word recognition (McClelland & Rumelhart, 1981), a model in which there are three levels of representation: a visual feature level, an abstract letter level, and an orthographic lexicon of local representations. They suggest that the locus of the weak activation could be damage to the letter feature level, or damage to the connections from feature level to letter level.

MNP also offer an interpretation of LBL reading in relation to this model of visual word recognition, and indeed provide some simulation results using the computational model of Jacobs and Grainger (1992), a model equivalent to the IAM except that it applies to five-letter words rather than four-letter words. MNP chose a specific lesion locus in the model: the excitatory connections from the word level back to the letter level, which they lesioned by setting the connection strengths to zero. They report that this exaggerates the effect of frequency on the model's naming latency, and point out that their patient had an exaggerated frequency effect on probability that a word would be read correctly.

Note, however, that the locus of lesion they chose was neither of the loci suggested by BPN, and it is not clear that the lesion locus choice by MNP was a felicitous one. We must not lose sight of the fact that LBL readers read nonwords in letter-by-letter fashion too. It is by no means clear how a lesion of connections to or from an orthographic lexicon could cause both words and nonwords to be read in LBL fashion, whereas a lesion before the letter level (as proposed by BPN), or a lesion at the letter level, should affect both words and nonwords.

What Causes the Patients to Read Letter-by-letter?

Any claim about the nature and locus of the impairment within some functional model of reading that is responsible for the impaired reading in this disorder must offer some explanation of why the abnormal reading takes the specific form it does, i.e. reading letter-by-letter. Several such explanations are offered.

SC suggest that, because of the left-hemisphere damage, the letters of a stimulus, represented in processed form (such as abstract letter identities generated in the right hemisphere), have to be transmitted across the corpus callosum from the right hemisphere to a language production centre in the left hemisphere; the letter-by-letter reading behaviour arises because that transfer process is slow, error-prone and serial.

CC and MC suggest that, because of the left-hemisphere damage, the letters, represented in raw unprocessed form, have to be transmitted across the corpus callosum from the right hemisphere to be recognised by an orthographic system in the left hemisphere, and that transfer process is slow, error-prone and serial.

BPN propose that, because letter activation is abnormally weak, individual letter identification requires amplification of activation, by focal attention or even fixation. So the letters in a letter string can no longer be identified in parallel, as normally happens in visual word recognition, but have to be sequentially attended to—maybe even individually fixated. Fixations occur from left to right, and it is plausible that sequential allocation of attention would too. Hence reading is not only letter-by-letter, but also left-to-right.

What Contribution Does the Right Hemisphere Make?

MC and CC do not consider that the right hemisphere contributes much, if anything, that is linguistic in the behaviour of LBL readers. That hemisphere is merely a source of raw visual information transmitted across the corpus callosum to the left hemisphere, where linguistic processing begins.

SC have a different view. Firstly, the right hemisphere is capable of identifying letters (so that what it sends across the callosum has already been orthographically processed). Secondly, the right hemisphere is capable of recognising and comprehending printed words to some degree, especially when these are concrete words. Thus it can perform lexical decision and semantic categorisation to some degree, and these capacities are responsible for the covert or implicit lexical and semantic processing reported in cases of LBL reading.

BPN offer yet another view, according to which the interactive activation system which subserves reading is distributed across both hemispheres in normals, and remains so in LBL readers. Like normal readers, LBL readers draw on resources from both hemispheres, differing only from normal readers in that the lexical activations evoked in this bihemispheric reading system are abnormally weak.

Why is LBL Reading Accuracy Affected by Imageability (Concreteness)?

According to SC, this is a contribution from right-hemisphere word-recognition and word-comprehension systems, which process high-imageable (i.e. concrete) words better than abstract words. On this view, there should be no effect of imageability on reading accuracy of LBL readers when they are reading aloud in LBL mode, a mode which does not exploit right-hemisphere lexical and semantic competences. However, the data in Figure 1 of the BPN paper document effects of imageability on reading accuracy in some, though not all, LBL readers when they are reading aloud.

According to BPN, the concreteness effect reflects a property of the intact reading system, namely that semantic representations are richer for concrete than for abstract words. When activation levels are weak and top-down help is greatly needed by the orthographic levels, such help is more strongly provided by semantics when a word is concrete than when it is abstract.

Why Do Some Patients Show Covert Processing and Others Not?

In a number of studies using exposure durations too brief to allow correct report of words, and indeed so brief that the patient might complain of having seen nothing at all, performance on lexical decision tasks and on semantic categorisation tasks has been found to be well above chance, and letter report in the Reicher-Wheeler task is better when the stimulus is a word than when it is a nonword. However, not all studies which have sought such effects have observed them. Why not? Several answers have been suggested:

One factor suggested by SC is that, if these covert-processing effects depend upon the right hemisphere, and if there are substantial individual differences in the linguistic capabilities of different right hemispheres, with the right hemispheres of some individuals have little or no language, then this kind of variability across LBL patients is to be expected.

A second factor suggested by SC is that one has to do these covert-processing experiments correctly. The LBL patient must not be using the LBL strategy since that overrides any output that the right hemisphere might be proposing. To minimise the likelihood that this strategy will be used, it is necessary to present the stimuli at brief enough durations that the strategy never yields word identification, to avoid asking for overt reading responses as well as categorisation responses, and to encourage the patient to guess.

The BPN view is that the activation level needed to support lexical decision or semantic categorisation is lower than the level needed to support overt reading aloud (which is why some LBL patients exhibit covert-processing effects), but in severe cases of LBL reading activation is so reduced that it cannot support even covert-processing tasks.

CONCLUSIONS

Although there is currently no agreement as to the answer to any one of these seven questions about pure alexia, the development of these questions out of the work reported in the papers in this volume has made it very clear what are the critical things we need to find out next about pure alexia. We need eye-movement studies of the kind alluded to by BPN to tell us more about why the letter-by-letter reading behaviour occurs in this disorder. Functional imaging or transcranial magnetic stimulation studies should provide crucial evidence concerning the part played by the right hemisphere. Scrupulous and detailed investigations of visual perceptual processing with nonlinguistic stimuli should be part of every study of pure alexia if we are to decide whether it is a reading-specific disorder or a more general visual disorder. And we need to get to the bottom of the issue of why some of these patients reveal covert knowledge of printed words before they can read them aloud whilst others do not.

REFERENCES

Déjerine, J. (1891). Sur un cas de cécité verbale avec agraphie suivi d'un autopsie. *Mémoires de la Société Biologique, 3*, 197–201.

Farah, M.J., & Wallace, M.A. (1991). Pure alexia as a visual impairment: A reconsideration. *Cognitive Neuropsychology, 8*, 313–334.

Jacobs, A.M., & Grainger, J. (1992). Testing a semi-stochastic variant of the interactive activation model in different word recognition experiments. *Journal of Experimental Psychology: Human Perception and Performance, 18*, 1174–1188.

Marshall, J.C., & Newcombe, F. (1973). Patterns of paralexia. *Journal of Psycholinguistic Research, 2*, 175–199.

McClelland, J.L., & Rumelhart, D.E. (1981). An interactive activation model of context effects in letter processing. Part I: An account of basic findings. *Psychological Review, 88*, 375–407.

A LITERATURE REVIEW AND NEW DATA SUPPORTING AN INTERACTIVE ACCOUNT OF LETTER-BY-LETTER READING

Marlene Behrmann, David C. Plaut and James Nelson
Carnegie Mellon University, Pittsburgh, USA

We present a theoretical account of letter-by-letter reading (LBL) that reconciles discrepant findings associated with this form of acquired dyslexia. We claim that LBL reading is caused by a deficit that affects the normal activation of the orthographic representation of the stimulus. In spite of this lower-level deficit, the degraded orthographic information may be processed further, and lexical, semantic, and higher-order orthographic information may still influence the reading patterns of these patients. In support of our position, we present a review of 57 published cases of LBL reading in which we demonstrate that a peripheral deficit was evident in almost all of the patients and that, simultaneously, strong effects of lexical/semantic variables were observed on reading performance. We then go on to report findings from an empirical analysis of seven LBL readers in whom we document the joint effects of lexical variables (word frequency and imageability) and word length on naming latency. We argue that the reading performance of these patients reflects the residual functioning of the same interactive system that supported normal reading premorbidly.

INTRODUCTION

Letter-by-letter (LBL) reading is the term used to define the reading pattern of premorbidly literate patients who, following brain damage acquired in adulthood, take an abnormally long time to read even single words. This reading deficit is typically associated with a lesion in the posterior portion of the dominant hemisphere and sometimes, but not always, accom-

---

Requests for reprints to Marlene Behrmann, Department of Psychology, Carnegie Mellon University, Pittsburgh, PA 15213-3890, USA (E-mail: behrmann+@cmu.edu).

This work was supported by NIMH FIRST awards to MB (MH54246-01) and to DCP (MH55628-01). We thank our colleagues, Dr. Sandra Black, Dr. Graham Ratcliff, and Dr. Jason Barton for referring the patients to us. We also thank Max Coltheart, Martha Farah, Marcel Kinsbourne, Jay McClelland, Marie Montant, Tatjana Nazir, Eleanor Saffran, Tim Shallice, and an anonymous reviewer for their insightful and constructive suggestions.

panied by a lesion of white matter tracts such as the splenium of the corpus callosum or forceps major (Damasio & Damasio, 1983). The reading performance of such patients is characterised by a "word length effect", a significant increase in naming latency as a function of the number of letters in a string. Times in the order of 1–3 seconds per additional letter in a string have been measured for some LBL readers, although there is considerable variability in reading speed across patients (Hanley & Kay, 1996). When the reading deficit manifests in the absence of other reading, writing, or spelling deficits, it is referred to as "pure alexia". When other written language deficits do accompany the LBL reading, they usually consist of surface dyslexia (Bowers, Bub, & Arguin, 1996; Friedman & Hadley, 1992; Patterson & Kay, 1982) or surface dysgraphia (Behrmann & McLeod, 1995; Rapp & Caramazza, 1991). Although less frequent, there are also reports of at least one case of deep dyslexia (Buxbaum & Coslett, 1996) and one case of phonological dyslexia (Friedman et al., 1993; Nitzberg-Lott, Friedman, & Linebaugh, 1994) accompanying LBL reading.

Two critical empirical findings concerning LBL reading pose difficult challenges for theories of this disorder. One finding is that these patients are impaired at letter processing. A second important finding is that some of these patients have available to them lexical and semantic information about the stimulus, as evidenced in their above-chance performance on lexical decision and semantic categorisation tasks. This latter finding suggests that the visual stimulus has been processed to a sufficient extent to produce such higher-order or later effects on performance. Two classes of theory have emerged, each of which emphasises one of these two paradoxical findings. One class argues that the deficit occurs early in processing, prior to the activation of an orthographic representation, and the early visual deficit observed in LBL readers is consistent with this view. We will refer to such views as *peripheral*, consistent with the Shallice and Warrington distinction in which impairments that adversely affect the attainment of the visual word-form are considered to be peripheral (Shallice, 1988; Shallice & Warrington, 1980). By contrast, if the impairment is at a later stage, the dyslexia is classified as *central*. With respect to LBL reading, central views maintain that the deficit occurs after the activation of a well-specified orthographic description and thus lexical and semantic information can still be accessed.

What is probably evident, even from this brief description, is that the peripheral and central accounts of LBL readings are difficult to reconcile. Peripheral views cannot readily account for the lexical/semantic findings and the central views do not explain the impaired early visual processing in these patients. In the present paper, we will argue that both sets of empirical findings can be accommodated within a single, interactive reading system to which both hemispheres contribute. We first present in detail the findings for a low-level deficit in LBL readers and then we consider the details of the lexical and semantic findings. Thereafter, we examine not only the peripheral and central accounts but also views that incorporate some aspects of both. Finally, we present our account and, to substantiate it, we review most of the published cases of LBL reading and present new empirical data from seven LBL readers.

KEY EMPIRICAL FINDINGS

Impairment in Early Visual Processing

It is now well established that patients with LBL reading do not activate orthographic representations adequately; for example, many patients are adversely influenced by alterations of the surface characteristics of the stimulus (for example, poorer reading of script than print) and many make mostly visual errors in reading (for example, JAY → "joy"). There are, however, several different explanations for the impairment in activating orthographic representations. Some studies have claimed, for example, that these patients suffer from a general perceptual deficit that impairs all forms of visual processing (Behrmann, Nelson & Sekuler, 1998; Farah, 1991; Farah & Wallace, 1991; Friedman & Alexander, 1984; Sekuler & Behrmann, 1996), although letter and word recognition might perhaps be especially vulnerable (Farah, 1997). Others have claimed that the impairment is specific to orthography and impairs the identification of letters per se (Arguin & Bub, 1993; Karanth, 1985; Kay & Hanley, 1991; Reuter-Lorenz & Brunn, 1990), or affects the rapid processing of multiple forms in parallel (Kinsbourne & Warrington, 1962; Levine & Calvanio, 1978; Patterson & Kay, 1982; Schacter, Rapcsak, Rubens, Tharan, & Laguna, 1990; see also Miozzo & Caramazza, this issue). Still others have maintained that LBL reading arises from insufficient attentional resources, thus forcing a serial, left–right strategy (Rapp & Caramazza, 1991), or from a problem in capacity or in switching visual attention (Buxbaum & Coslett, 1996; Price & Humphreys, 1992). In a recent paper, one of us (Behrmann & Shallice, 1995) argued that the core deficit is one of letter processing and that the time to activate the representation for even a single letter is slow. Importantly, we argued that the processing deficit is not spatially determined, i.e. processing does not proceed from left to right of the array; rather, it has to do with serial order such that letters appearing later in a string (even when all letters are presented at fixation in an RSVP paradigm) are disadvantaged relative to letters appearing earlier in the string.

It is important to note that many of these peripheral views are not necessarily mutually exclusive. For example, although the critical deficit may be one of disordered perceptual processing, an obvious consequence of such a deficit is an impairment in rapid and accurate letter processing and identification (Behrmann & Shallice,1995; Sekuler & Behrmann, 1996). What is central to all these peripheral views (labelled under the heading orthographic access view, [Bowers et al., 1996]), however, is that the fundamental impairment is one of visual processing, arising relatively early and preventing the derivation of an adequate orthographic representation.

Numerous investigators have commented that problems in letter-processing tasks are so common in LBL reading that impaired letter identification is likely to underlie their reading deficit (Coltheart, 1981; Patterson & Kay, 1982). In support of this, Patterson & Kay showed that all four of their patients made letter identification errors, albeit to varying degrees. Patient CH, for example, identified correctly only 16/26 upper-case and 10/26 lower-case letters and chose the odd letter out (for example, f F E) correctly on only 75% of the trials. Patient TP, on the other hand, identified 25/26 lower-case letters and made no errors on cross-case matching (e.g. D d). Interestingly, the types of errors made by all the patients reflected the visual similarity between the target and the error, suggesting that visual feature overlap is a major factor in letter misidentification (see also Perri, Bartolomeo, & Silveri, 1996). Consistent with the findings of letter misidentification, Behrmann and Shallice (1995) maintained that there is no convincing evidence of normal letter processing in any LBL reader. They then posed the challenge that, unless the hypothesis of impaired letter processing was found wanting, this should constitute the default explanation for the reading impairment in patients with LBL reading.

Lexical and Semantic Effects on Reading

Give this evidence, it is particularly puzzling that some LBL readers, even if they cannot explicitly identify the stimulus, can nonetheless demonstrate some lexical and semantic information about it. This result was documented in several fairly early reports of LBL patients (Albert, Yamadori, Gardner, & Howes, 1973; Caplan & Hedley-Whyte, 1974; Kreindler & Ionescu, 1961); even though the patients in these studies could not identify a written word overtly, they were still able to match this target with a word spoken by the investigator, or with a visually presented object. These initial observations of implicit or covert reading abilities of LBL readers in the absence of explicit identification have been upheld in a number of more recent studies. For example, Shallice and Saffran (1986) reported that their patient, ML, was above chance at performing lexical decision and semantic categorisation tasks with stimuli presented too briefly to permit overt identification. Several other studies have similarly shown that their patients can perform lexical decision (Bub & Arguin, 1995) as well as semantic classification of words (for example, living vs. nonliving) at exposure durations too brief for the patients to have identified the target items explicitly (for example, Bub & Arguin, 1995; Howard, 1991). In the largest series, Coslett and colleagues (Coslett & Saffran, 1989a, 1994; Coslett, Saffran, Greenbaum, & Schwartz, 1993) have described five patients who performed well above chance on lexical decision and semantic categorisation tasks with the very same stimuli they could not identify explicitly (see also Saffran & Coslett, this volume). Coslett and colleagues also described two additional patients who fit the definition of optic aphasia and who were completely unable to name letters (or any other stimuli from visual presentation). Even in the absence of letter naming, these two patients were fairly successful at lexical decision and binary categorisation tasks (Coslett & Saffran, 1989b, 1992).

THEORIES OF LETTER-BY-LETTER READING

The two empirical findings of an early deficit and the influence of later lexical and semantic properties of the stimulus on performance are now both well documented in the domain of LBL reading. As mentioned previously, there are two main classes of explanation of LBL reading which differ in the extent to which they emphasise one or the other of these two findings. The peripheral view argues that the deficit occurs early in processing, consistent with the early visual deficit observed in these readers. Proponents of this view have focused on the letter-processing deficit and its underlying mechanism without paying much attention to the later, lexical and semantic effects. The central view argues that the deficit occurs only after (or at least does not prevent) the activation of a well-specified orthographic description and, thus, lexical and semantic information can still be assessed. Within the central view there are two different accounts. One account, although perhaps the less favoured at present, is that the patients have an intact reading system and can activate lexical and semantic knowledge normally, but that the output of this intact system is disconnected from consciousness (Schacter, McAndrews, & Moscovitch, 1988; Young & Haan, 1990). Thus, although subjects can process the information, and the results may be revealed through implicit tasks that do not require explicit processing, such as semantic categorisation or priming, the contents of this system are not available for overt report. This view has probably fallen out of favour for a number of reasons, including the fact that it cannot account for the existence of the early visual processing impairment and that there is no explanation for the hallmark feature of this problem, the letter-by-letter reading itself or the increase in naming latency with word length.

A second central view argues for a visual-verbal disconnection, i.e., that the visual areas involved in reading are anatomically and/or functionally disconnected from the more semantic/conceptual areas. The best example of this view is from Déjerine who, in his famous 1891/1892 case studies (for overview, see Bub, Arguin, & Lecours, 1993), interpreted the LBL syndrome as a disconnection of the visual verbal input from "the visual memory centre for words", which is located in the left angular gyrus (Geschwind, 1965; Greenblatt, 1973; Speedie, Rothi, & Heilman, 1982). Bowers et al. (1996a; also Arguin, Bub, & Bowers, this issue) have adopted a similar perspective and maintained that the disconnection arises only after orthographic word representations have been activated; thus the disconnection is between orthographic representations (logogens) on the one hand and phonological codes on the other. This disconnection delays (or precludes) access to the phonological code for naming, while leaving lexical decision, semantic categorisation, and the word superiority effect intact. In support of their disconnection account, they showed that the reading performance of a LBL reader, IH, was facilitated by orthographically related primes (e.g. GATE-gate) but, unlike normal readers, not by homophonically related primes (e.g. gait-gate) (Arguin, Bub, & Bowers, this issue). IH also showed no effect of phonemic neighbourhood size on word recognition performance. Given that IH shows no evidence for covert phonological activation, the functional site of the lesion, according to this account, lies between orthographic and phonological processing (but see Montant, Behrmann, & Nazir, 1998, for demonstration of phonological priming in an LBL reader).

In summary, both the peripheral and central explanations of LBL reading do well at accounting for the subset of empirical findings on which they focus. The limitations of most of these accounts, however, is that they do not simultaneously accommodate both the peripheral and central aspects of LBL reading. One exception to this is the view of Coslett and Saffran, which takes into account both of these aspects (Buxbaum & Coslett, 1996; Coslett & Saffran, 1994; Saffran & Coslett, this issue). These authors have suggested that there is a deficit (most likely peripheral and early) in the normal reading system in these patients and that the sequential output pattern typically associated with LBL reading is mediated by the left hemisphere. The covert reading and later lexical/semantic effects do not depend on the inaccessible (or degraded) word form generated by the left hemisphere. Rather, a separate reading mechanism, subserved by the intact right hemisphere, is responsible for the findings of preserved lexical decision and semantic categorisation. On this *right-hemisphere* interpretation, the lesion in pure alexia prevents visual information from both cortices from accessing lexical and semantic representations in the left hemisphere, but this information can still support the (albeit limited) reading capabilities of the right hemisphere. Consistent with the right-hemisphere view, Coslett et al. (1993) showed that, as the LBL patients became able to engage in covert reading, explicit recognition performance was influenced by word imageability and grammatical class, properties often associated with the reading skills of the right hemisphere (Coltheart, 1980, 1983).

A particularly appealing aspect of this right-hemisphere view is that the later lexical and semantic effects observed in LBL readers are similar to those generated by the right hemisphere of commissurotomy patients as well as by left-hemispherectomy patients (Coltheart, 1983). In fact, a similar explanation has been proposed by some to account for the pattern of reading in patients with deep dyslexia (Coltheart, 1980, 1983; Saffran, Bogyo, Schwartz, & Marin, 1980).

Although the right-hemisphere account of LBL reading takes into account both its peripheral and central aspects, it makes the very specific assumption that these different aspects reflect entirely distinct modes of reading and are mediated by different hemispheres. This assumption would seem to imply that the normal modes of operation of the two hemispheres are altered in LBL reading, with the right hemisphere playing an increased role relative to its normal contribution. However, the exact relationship between the roles of the left and right hemisphere in LBL reading and their roles in normal reading has never been made explicit on the right-hemisphere account. By contrast, we present an alternative theory in which both the early and late effects in LBL reading arise as a result of a peripheral impairment to the normal word reading system that is supported by both hemispheres.

AN INTERACTIVE ACCOUNT OF LETTER-BY-LETTER READING

One possible reaction to the controversy between the peripheral and central accounts of LBL reading is to argue that reconciliation is unnecessary and that a host of different underlying deficits may give rise to this problem. Moreover, individual differences could also arise from the different compensatory strategies adopted by the patients (Price & Humphreys, 1992, 1995). On this view, no uniform account of letter-by-letter reading exists and none is needed. In this paper, however, we argue that a reconciliation between these various positions is both desirable and possible and that LBL may be accounted for by a single, unifying view of the normal word reading system.

Our account takes as its starting point the framework of the Interactive Activation Model of letter and word perception (hereafter IAM; McClelland & Rumelhart, 1981; Rumelhart & McClelland, 1982). The model contains three levels of processing units—letter features, letters, and words—such that elements and their components (e.g. the word MAKE and the letter M in the first position) mutually support each other whereas inconsistent alternatives (e.g. the words MAKE and TAKE) mutually inhibit each other. The critical properties of the model for our purposes are that processing is *cascaded* and *interactive*. Cascaded processing means that partial results at each level, in the form of intermediate levels of activation, propagate to other levels immediately and continuously, rather than waiting until processing at lower levels is complete. Interactive processing means that activation not only propagates from lower to higher levels, but that the activation at higher levels feeds back to lower levels to provide additional support for those lower-level elements that are consistent with the higher-level activation. Thus, cascaded, interactive processing causes early letter activation to feed forward and partially engage word representations, which in turn feed back to the letter level to influence subsequent processing.

We should point out that we are adopting the IAM as a framework for explaining LBL reading not out of any commitment to the localist word representations it contains, but because it provides a clear instantiation of the principles of cascaded, interactive processing. The same principles apply within distributed accounts of lexical processing (e.g. Plaut, McClelland, Seidenberg, & Patterson, 1996; Seidenberg & McClelland, 1989; Van Orden, Pennington, & Stone, 1990) and, as we explicate below, essentially the same account of LBL reading holds within such models.

We claim that the fundamental impairment in LBL reading, following an occipital lesion, is a general perceptual deficit that degrades the quality of visual input. In the IAM, this deficit can be conceptualised as damage to the letter feature level or between this level and the letter level. The impact of this perceptual impairment on word recognition is that it permits only weak or partial parallel activation of the letters in a word. This weak activation does not suffice for explicit identification of the word (i.e. no word unit achieves a sufficiently high level of activation to exceed the response threshold) and the system must resort to sequential processing to enhance the activation of individual letters. Critically, this type of sequential processing is not an abnormal strategy only employed following brain damage, but is the manifestation of the normal reading strategy of making additional fixations when encountering difficulty in reading text (Just & Carpenter, 1987; Rayner & Pollatsek, 1987). For example, normal subjects fixate more frequently in a long word than in a short word in order to enhance the quality of the stimulus (O'Regan & Levy-Schoen, 1989). LBL readers also fixate more frequently; in fact, given the very poor quality of the visual input, they fixate almost every letter (Behrmann, Barton, & Black, 1998), giving rise to the hallmark word length effect. Presumably these fixations aid performance by permitting the increased spatial resolution of the fovea to be applied to multiple locations within the word. In fact, even in the absence of overt saccades, a word length effect would be expected, given that LBL readers can improve perceptual processing by rescaling covert attention from the entire word to apply successively to letters within the word[1].

Even though the word-level activation produced by the initial, weak, parallel letter activation is insufficient to support explicit identification, it nonetheless would be expected to activate the correct word to a greater extent than its competitors, and to produce more activation overall than would be produced by a nonword (see McClelland & Rumelhart, 1981). Thus, assuming this lexical activation propagates further into a semantic system (not implemented in the IAM, but see McClelland, 1987), during the course of the sequential processing, activation from individual letters propagates into the system and adds to the cumulative activation at the word level. Concurrently, this word-level activation feeds back to the letter level to facilitate subsequent recognition of the word's letters. The strength of this top-down support is a function of the degree of partial world-level activation. We assume that, in a more general system including semantics, the degree of higher-level activation would scale not only with frequency (as in the IAM) but also with imageability, such that words of higher frequency or imageability would be more active, and hence provide stronger top-down support for letter activations, than would words of lower frequency or imageability. Consequently, the system would

---

[1] Although the IAM does not incorporate mechanisms for overt or covert shifts of attention, the model could be extended to include such a mechanism (see, e.g. Hinton, 1990), and other word reading models that are consistent with cascaded, interactive processing do contain such mechanisms (e.g. Plaut, McClelland, & Seidenberg, 1995)

converge more quickly on responses for high-imageability and high-frequency items compared with low-frequency and low-imageability items.

A further determinant of the degree of the strength of top-down support is the time over which higher-level activation can accumulate, and it is this time factor that gives rise to an interaction between these lexical variables and word length. Given that longer words take longer to process in LBL reading, their higher-level representations have longer to integrate bottom-up support and, therefore, produce stronger top-down effects on performance. Thus, there is more opportunity for frequency and imageability to influence the recognition of seven-letter words compared with three-letter words. Indeed, even in normal subjects, there is a significant interaction of word length and frequency; Weekes (1997) found that, whereas, there was no effect of word length on the naming latency of high-frequency words, there was a small effect for low-frequency words and a more marked effect for nonwords.

The essence of our account, then, is that LBL readers make use of the same cascaded, interactive system as normal readers but are prompted to resort to sequential processing more often (manifest either as multiple eye movements or shift of covert attention) to compensate for the degradation in visual input following the left occipital lesion. Nonetheless, the weak, parallel activation from this input propagates to higher levels of the system to engage lexical/semantic representations partially, and these representations provide top-down support that facilitates subsequent lower-level processing. This account provides a unified explanation of both the early visual and later lexical and semantic findings in LBL reading.

In fact, an interactive theory of this general form has already been proposed to account for another type of acquired peripheral dyslexia, neglect dyslexia (Behrmann, Moscovitch, Black, & Mozer, 1990; Mozer & Behrmann, 1990). According to this theory, neglect dyslexia arises from a low-level deficit in spatial attention that affects the bottom-up processing of one-side (typically the left) of a visual stimulus. Given that reduced attention does not completely filter out information, but only lowers the likelihood that it is propagated into the system, the unattended information on the neglected side of the stimulus may still engage higher-level processes, albeit to a lesser extent than the non-neglected information. Provided that the attentional deficit is not too severe, stimulus information from the neglected side may be processed sufficiently to engage higher-level (lexical / semantic) representations. Thus, factors such as lexical status and morphological composition can still influence performance, leading to, for example, better reading of words compared with nonwords and better reading of compound words compared with two single, unrelated words (Behrmann et al., 1990; Brunn & Farah, 1991; Làdavas, Umiltà, & Mapelli, 1997; Sieroff, Pollatsek, & Posner, 1988). Thus, as in LBL reading, both "early" effects (e.g. influence of stimulus size, horizontal position, and word length) and "late" effects (e.g. influences of lexical status and morphology) may be observed in the same patient.

To this point, we have cast our views in terms of the IAM, in which the critical lexical effects are mediated by word-specific (localist) processing units. However, as alluded to earlier, this choice is to a large extent merely expository. The same lexical effects can be recast within models of lexical processing employing distributed representations (e.g. Seidenberg & McClelland, 1989; Plaut et al., 1996; Van Orden et al., 1990). In a distributed representation, words are encoded by distinct but overlapping patterns of activity, such that each word is represented by the activity of many units and each unit participates in representing many words. Within a distributed system, lexical knowledge—the fact that certain patterns of activity but not others correspond to familiar stimuli—is reflected not in the structure of the system but in the dynamics of how the units interact. Specifically, lexical knowledge in the form of learned weights on the connections between units causes the activity pattern corresponding to each word to form an *attractor*. What this means is that, when the system is in an unfamiliar pattern of activity—one that does not correspond to a known word—interactions among units alter this pattern so that it moves towards and ultimately settles to the nearest (most similar) familiar attractor pattern. A critical property of attractor dynamics for our purposes is that they can be partial; an activity pattern that is sufficiently far from the nearest attractor may still be pulled towards it (thereby reflecting its lexical properties to a degree) but may not settle it completely (which would correspond to explicit recognition). Thus, the partial activation and competition among word units in the localist IAM corresponds to the partial movement of activity patterns towards word attractors within distributed systems. Yet both types of systems employ cascaded, interactive processing such that partially degraded orthographic input is propagated throughout the system to engage these "lexical" effects to varying degrees. Thus, we claim that our account of LBL within the framework of the IAM also holds for distributed connectionist models of lexical processing.

In fact, many of the effects that we ascribe to top-down activation from word units in the IAM have been demonstrated in existing simulations of distributed attractor networks. For example, Hinton and Shallice (1991) demonstrated relative sparing of categorisation performance with poor explicit identification following lower-level damage to an attractor network that was trained to map orthography to semantics. Plaut and Shallice (1993) replicated this finding in an attractor network that mapped orthography to phonology via semantics, and also showed that, for the words the network failed to read correctly following damage near orthography, it was nonetheless well above chance ($d' = 1.80$)[2] at lexical decision for these words. Plaut and Shallice also demonstrated in a similar network that high-imageability words develop stronger semantic

---

[2] This d' value is reported incorrectly as 1.36 by Plaut and Shallice (1993, p. 445).

attractors than low-imageability words—as reflected by their relative robustness to damage—if imageability is instantiated in terms of the relative richness of a word's semantic representations (i.e. how many semantic features are accessed consistently across contexts; also see Plaut, 1995). The stronger semantic attractors for high-imageability words would naturally lead to stronger top-down effects on orthographic representations, although this was not tested directly. And finally, in conjunction with the interactive theory of neglect dyslexia mentioned earlier, Mozer and Behrmann (1990, 1992) simulated the co-occurrence of many of the lower-level perceptual effects and higher-level lexical/morphological effects observed in neglect dyslexia by damaging input to the attentional mechanism within MORSEL (Mozer, 1991), a network model that implements attractors for words using a "pull-out" network over letter-cluster units.

In contrast with the right-hemisphere account of LBL reading (Buxbaum & Coslett, 1996; Coslett & Saffran, 1994; Saffran & Coslett, this issue), our view does not specifically implicate a particular hemisphere as the locus of the lexical and semantic effects. We have argued that the simultaneous presence of visual and lexical/semantic findings arise from a single interactive system, involving both hemispheres, and that there is no compelling reason to invoke the separate right hemisphere as the sole (or even primary) mediator of the postlexical effects. Importantly, this view entails that, in LBL readers, the reading system per se is unchanged from its premorbid state; reading behaviour arises as a consequence of degraded input to the same cascaded, interactive systems that supported normal reading premorbidly. Although we do not question the fact that the right hemisphere has some language and reading abilities (Beeman & Chiarello, 1997; Michel, Henaff, & Intriligator, 1996; Vargha-Kadem et al., 1997), our claim is that the lexical and semantic findings in LBL readers do not arise solely from the more primitive right-hemisphere reading system. Rather, these higher-level effects emerge from the residual workings of the interactive reading system that involves both hemispheres, governed by the same computational principles.

What type of evidence would support our interactive account of LBL reading? First, all patients should have an early deficit of some form that prevents the rapid and reliable activation of orthographic information. Second, variables considered to be diagnostic of later lexical processes should also influence the reading performance of these patients. In support of our position, in this paper we present a review of the existing literature as well as new empirical data obtained from seven LBL readers. As the findings will indicate, both the literature review and the empirical data are consistent with our position, showing that early visual deficits are identifiable in almost every patient (and not demonstrably absent in the remaining cases) and that, when they have been investigated carefully, strong effects of lexical variables are also observed. That a partial low-level deficit can permit partial higher-level activation (supporting lexical decision and categorisation performance) that further influences lower-level processing (producing frequency and imageability effects on reading latencies) is, on our view, a direct consequence of the cascaded, interactive nature of the normal reading system.

REVIEW OF PUBLISHED CASES OF LETTER-BY-LETTER READING

This review of the published studies of 57 LBL readers was undertaken both to document the existence of early deficits and to examine the extent to which there is higher-level processing in these patients. Although this review is intended to be as comprehensive as possible (see Table 1), we have deliberately excluded several papers. Many of the excluded papers have a different focus and do not provide sufficient detail for our analysis; for example, some papers focus on the anatomical aspects of the case rather than on the reading performance per se (Ajax, 1967; Damasio & Damasio, 1983; Greenblatt, 1973), others describe aspects of the patient's performance such as the intact reading of stenography (Regard, Landis, & Hess, 1985), an associated deficit in music reading (Horikoshi et al., 1997), or associated colour deficits (Freedman & Costa, 1992), which are unrelated to the issue at hand, and yet other papers report a rehabilitation procedure for the patient without including much detail on the patients' pre-therapy reading performance (Kashiwagi & Kashiwagi, 1989; Tuomainen & Laine, 1991). There are also a small number of papers that are not included simply because the description of the patient's reading is unclear or insufficiently detailed for our purposes (Caplan & Hedley-White, 1974; Kreindler & Ionescu, 1961), although we have generally made reference to these papers elsewhere in the text. We are also aware of posters of LBL readers presented at conferences and have not included those (with the exception of the Vigliocco, Semenza, & Neglia, 1992, because the description of the patients is sufficiently detailed). We have also come across a few papers published in other languages (El Alaoui-Faris et al., 1994), but have restricted our analysis to published English papers. Finally, to make the table as comprehensive as possible, we have included our own patients at the end. To do so, we have combined the findings from the empirical analyses we describe in the latter part of this paper with previous descriptions of these patients' reading performance.

Procedure

For this analysis, we reviewed the published papers to determine whether there was any evidence for a peripheral deficit in the patient. If the authors of a paper classified the patient's deficit as occurring early (peripherally or prelexically), this was counted as positive evidence. We also took as positive evidence a reported perceptual difficulty in single-letter processing or identification or a more general perceptual deficit in which other visuoperceptual abilities are impaired, even if the authors did not classify the patient in the peripheral group per se. We also determined whether there was evidence for lexical and semantic effects in these same LBL readers. This consisted of reviewing the reported papers and determining whether there was positive evidence for: (1) covert reading under brief exposure (either in lexical decision or semantic categorisation tasks), (2) a word superiority effect, (3) effects of frequency, imageability, regularity, and part-of-speech on naming words, and (4) effects of frequency and imageability on lexical decision. With regard to the word superiority effect, we simply noted the better report of items in words over nonwords in any experimental paradigm (not solely the standard Wheeler-Reicher type task) and did not take special note of a pseudoword effect (better report of items in legal pseudowords over illegal nonwords).

In analysing the effects of lexical variables such as frequency, imageability etc., we considered their impact on naming and on lexical decision under both brief and prolonged exposure duration. We have only documented whether these variables significantly influence the patient's reading performance (i.e. as a main effect) in either reaction time or accuracy. Very few studies present the interaction between these different variables and word length (see Doctor, Sartori, & Saling, 1990, for an exception) so we do not consider the interactions in our tabulation, although we consider it to be crucial for our interactive view. Finally, some studies (for example, Buxbaum & Coslett, 1996; Coslett & Monsul, 1994; Coslett & Saffran, 1989a; Coslett et al., 1993; Doctor et al., 1990; Friedman et al., 1993; Shallice & Saffran, 1986) assess the reading of other higher-order lexical, orthographic, or semantic variables, for example a comparison of performance on suffixed/pseudosuffixed words.

Although these other variables are relevant to the issue under investigation, reports of this were too few to warrant a separate category.

Because this analysis is retrospective, there are obviously a number of problems. In many cases, the critical variables are not tested (or perhaps tested but not reported or analysed statistically). Even when they are tested systematically, the dependent variable is often a measure of accuracy rather than reaction time. Yet the effects in LBL reading are typically more robust (and perhaps only evident) in reaction time, given the high degree of accuracy in many cases. We have not differentiated between the various dependent measures and simply note the presence of the main effect using either metric.

Results

Evidence for a Peripheral Deficit

Many researchers have previously observed that LBL readers invariably perform poorly on tests of letter recognition, although, to date, there has been no substantiation of this claim. Almost all the patients we reviewed had some noticeable problem with letter processing, as is evident from the first column of Table 1. Indeed, 50 of the 57 patients have a positive check mark in the first column. Although some of these patients were highly accurate on naming single letters for an unlimited exposure, a finding which might suggest no obvious impairment, the time to name the individual letters is often abnormally long. For example, DR (Doctor et al., 1990) correctly identified 24/26 lower-case and 25/26 upper-case letters, but

Table 1. Review of Published Cases of Letter-by-letter Reading

		Brief Exp					Naming				LD	
Patient	Study	PL	LD	SC	WSE	Fr	Im	Rg	PS	Fr	Im
DM	Arguin & Bub (1993, 1994a,b)	✓	✓	✓	✓	✓	–	–	–	✓	–
	Bub & Arguin (1995)	✓	✗	✗	–	✓	✓	–	–	–	–
EL	Behrmann et al. (1998)	✓	–	–	–	✓	–	✓	–	–	–
	Montant et al. (1998)										
IH	Bowers, Arguin, & Bub (1996)	✓	–	–	✓	✓	–	–	–	–	–
	Bowers, Bub, & Arguin (1996)										
	Arguin et al. (this issue)										
JV	Bub, Black, & Howell (1989)	✓	–	–	✓	✗	✓	–	✓	–	✗
n/a	Buxbaum & Coslett, (1996)	✓	✓	✓	–	✗	✓	✗	✓	✓	✗
1	Coslett & Saffran (1989a)	✓	✓	✓	–	+	✓	–	+	✓	✗
2	Coslett & Saffran (1989a)	✓	✓	✓	–	–	✓	–	✓	✓	–
3	Coslett & Saffran (1989a)	✓	✓	✓	–	✓	–	–	–	✓	–
4	Coslett & Saffran (1989a)	✓	✓	✓	–	–	✓	–	–	✓	–
JWC	Coslett et al. (1993)	✓	✓	✓	–	–	–	–	–	–	–
DR	Doctor et al. (1990)	✓	✓	✗	–	✗	✗	–	✗	–	–
RE	Feinberg et al. (1995)	✓	✓	✓	–	–	✗	–	✗	–	–
n/a	Friedman & Alexander (1984)	✓	–	–	✓	✓	✗	✓	✓	–	–
BL	Friedman & Hadley (1992)	?	–	–	–	–	✗	✗	✗	–	–
TL	Friedman et al. (1992)	✓	–	–	–	–	✓	✗	✓	–	–
n/a	Grossi et al. (1984)	✓	–	+	–	–	–	–	–	–	–
DC*	Hanley & Kay (1996)	✓	–	–	✓	–	✓	–	–	–	–
n/a	Henderson et al. (1995)	✓	–	–	–	–	✓	–	–	–	–
PM	Howard (1991)	✓	✗	✗	–	✓	✓	–	–	–	–
KW	Howard (1991)	✓	✗	–	–	✓	✓	–	–	–	–
PD	Kay & Hanley (1991)	✓	–	–	+	–	–	–	–	–	–
	Hanley & Kay (1992, 1996)				✗						
L	Kinsbourne & Warrington (1962)	✓	–	–	–	+	–	–	–	–	–
P	Kinsbourne & Warrington (1962)	✓	–	–	–	+	–	–	–	–	–
S	Kinsbourne & Warrington (1962)	✓	–	–	–	+	–	–	–	–	–
C	Kinsbourne & Warrington (1962)	✓	–	–	–	+	–	–	–	–	–
n/a	Landis et al. (1980)	✗	–	–	–	–	–	–	–	–	–
ON	Lazar & Scarisbrick (1993)	✓	–	–	–	–	–	✓	✓	–	–
CC	Levine & Calvanio (1978)	✓	–	–	✓	–	–	–	–	–	–

Patient	Reference									
RT	Levine & Calvanio (1978)	✓	–	–	–	–	–	–	–	–
GC	Levine & Calvanio (1978)	✓	–	–	–	–	–	–	–	–
CP#	Montant, Nazir, & Poncet (this issue)	✓	–	–	–	–	–	–	–	–
MW	Patterson & Kay (1982)	✓	✗	✗	–	–	–	–	–	–
CH	Patterson & Kay (1982)	✓	✗	✗	–	–	–	–	–	–
TP	Patterson & Kay (1982)	✓	✗	–	–	–	–	–	–	–
KC	Patterson & Kay (1982)	✓	–	✗	–	–	–	–	–	–
SP	Perri et al. (1996)	✓	✗	✗	✓	✓	–	–	–	–
EW	Price & Humphreys (1992)	✓	✗	✗	–	✓	+	–	–	–
HT	Price & Humphreys (1992)	✓	✗	–	–	+	✗	–	–	–
SA	Price & Humphreys (1995)	✓	✗	✗	–	–	–	–	–	–
WH*	Price & Humphreys (1995)	✓	–	✗	✓	✓	–	–	–	–
HR*	Rapp & Caramazza (1991)	✓	–	–	–	–	–	–	–	–
PT	Rapcsak et al. (1990)	?	–	–	–	–	✗	✗	–	–
WL	Schacter et al. (1990)	✓	–	–	✓	–	–	–	–	–
ML	Reuter-Lorenz & Brunn (1990)	✓	✓	✓	–	–	✗	–	✓	–
n/a	Shallice & Saffran (1986)	✓	–	–	–	–	–	–	–	–
BY	Stachowiak & Poeck (1976)	✓	–	–	✓	–	–	–	✓	–
BD	Staller et al. (1978)	?	✗	✗	–	–	–	–	–	–
RG	Vigliocco et al. (1992)	?	✗	✗	✓	–	–	–	–	–
RAV	Vigliocco et al. (1992)	?	–	–	✗	–	✗	–	–	–
JDC	Warrington & Shallice (1980)	?	–	–	–	–	–	–	–	–
Warrington & Shallice (1980)										

Our Patients

DS	Behrmann, Black & Bub (1990)	✓	–	–	✓	✓	–	–	–	–
	Behrmann & Shallice (1995)									
	Sekuler & Behrmann (1996)									
PC		✓	–	–	–	–	–	–	–	–
MW	Sekuler & Behrmann (1996)	✓	–	–	✓	✓	✗	–	–	–
DK		✓	+	+	–	✓	✓	–	–	–
MA	Sekuler & Behrmann (1996)	✓	–	–	✓	✓	✓	–	–	–
IS	Behrmann & McLeod (1995)	✓	–	✓	✓	✗	✗	–	–	–
TU	Farah & Wallace (1991)	✓	–	–	–	✓	✓	✗	–	–
	Sekuler & Behrmann (1996)									

PL = Prelexical; LD = Lexical Decision; SC = Semantic Category; WSE = Word Superiority Effect; Fr = Frequency; Im = Imageability; Rg = Regularity; PS = Part of Speech; ✓ = evidence for; ✗ = evidence against; + = some evidence for; ? = unclear; – = not tested; * = left-handed; # = ambidextrous.

the time to do so was 2.03 sec and 1.75 sec, respectively. Patients PD and DC (Hanley and Kay, 1996; Kay & Hanley, 1991) also made a relatively small number of letter misidentifications (26/26 and 25/26 lower-case letters, respectively) but their speed of processing was dramatically slowed relative to a control subject. This slowing was particularly evident when the two patients performed a letter-matching task and made same/different judgements on two letters (e.g. aR) presented simultaneously or sequentially (500msec SOA). The control subject showed a slight reaction time advantage for the sequential over the simultaneous condition. PD's reaction times were similar to the control in the sequential condition but he took roughly 400msec longer when the letters were presented simultaneously. DC was even slower than PD on the simultaneous condition but, in addition, was far slower than PD or the control subject on the sequential condition (see Hanley & Kay, 1996, Fig. 4). These findings suggest that neither of these patients process letters normally.

There remain seven contentious cases for whom a letter processing deficit is not clearly apparent. We consider each in turn. Patient BL scored 25/26 correct on upper-case naming but we have no indication of the time required to do so (Friedman & Hadley, 1992). There is also no report of other letter processing tests nor tests of general visual processing for BL. We do know, however, that BL showed a very steep serial position curve in reporting letters from four-item strings under tachistoscopic presentation. Whereas performance was reasonably accurate for items in position 1 (almost at 100% on word stimuli), performance dropped below 50% for items in position 4 of words (and close to floor for items in position 4 of nonwords). A similar account of preserved accuracy but abnormal speed of processing may account for the patient reported by Landis et al. (1980). Although this young man accurately identified single letters, this was done so hesitantly, especially when the letters were intermixed with numbers (see similar observation by Polk and Farah, reported in Farah, 1997). Both BD and GR (Vigliocco et al., 1992) performed similarly to normal control subjects on a letter-matching task under simultaneous and sequential conditions, suggesting that speed is normal for them. We do not, however, know what their accuracy was on this task. When presented with single letters for 2 sec on a separate task, a duration that is extremely long for normal subjects, BD identified 88% correctly and GR 90%. The errors produced were all visual confusions with the target (for example, d/b, v/u). It appears then that even these two subjects might not have normal letter processing and that the apparently normal reaction times may be due to a speed-accuracy trade-off.

Both RAV and JDC (Warrington & Shallice, 1980) appear to identify single letters particularly well, even when separated by either alphanumeric characters (Expt. 1) or when flanked by distractors (Expt. 2). Nonetheless, both patients are more affected than normal readers by items presented in script compared with print. Moreover, RAV appear to show an interaction in reaction times between script/print and word length. For three-letter words, the difference between script/print is 2.4 sec, whereas the difference increases to 5 sec for seven-letter words. Both the differences between script and print and the interaction with word length have been taken to be strong indicators of an early visual processing deficit (Farah & Wallace, 1991). Lastly, PT (Rapcsak, Rubens, & Laguna, 1990; Schacter et al., 1990) performed well on tasks of letter discrimination (letters vs. nonletters or mirror-reversed letters; 100%), cross-case matching (95%), and pointing to letters (100%). Although accuracy was low for letter naming, with a score of only 15/26 for upper-case and 15/26 for lower-case letters, this might possibly be a result of an anomia rather than of a letter recognition deficit per se. Once again, however, we have no indication that PT performs these simple letter tasks at normal speeds. Whether her letter processing is indeed normal, therefore, remains unanswered at present.

The findings from the review thus far suggests that there is no single subject for whom letter recognition is definitively normal. As is evident, patients may be impaired in speed, even if not in accuracy, and sometimes the converse is true. Additionally, as noted by Patterson and Kay (1982), even high accuracy might not be a satisfactory indicator of letter processing skill; an accuracy score of 85% in letter naming may not seem too serious an impairment but this might become very debilitating in a more taxing task when the subject is required to represent letters for the purposes of word recognition.

An important question that remains, then, is what are the consequences of this peripheral deficit for reading. We have assumed that the letter processing deficit plays a causal role in LBL reading and that, because the weakened activation is insufficient, subjects make both overt and covert gaze shifts to enhance the letter activation. A correlation between reading speed and the accuracy of single-letter identification, suggestive of a strong relationship between the two, was described by Shallice (1988) in a small group of eight LBL readers. This correlation, however, does not seem to be perfectly upheld, as a recent study described two patients with fairly similar letter recognition patterns but with very different performance in word recognition (Hanley & Kay, 1996). The exact relationship between poor letter and word recognition processes is beyond the scope of this paper. Also beyond the scope of this paper is whether the poor letter processing derives from an even more fundamental perceptual problem, as we (amongst others) have argued elsewhere (Behrmann et al., 1998; Sekuler & Behrmann, 1996; see also Chialant & Caramazza, this issue), or whether it is restricted to alphanumeric stimuli. For the present purposes, we are only interested in establishing that, across the population of LBL readers, there is strong evidence for a peripheral deficit that adversely affects letter processing. We now turn to the question of whether variables associated with later stages of word recognition are also observed in this population.

Evidence for Lexical and Semantic Effects

The major problem we encountered in this review is that very few studies systematically document the higher-level effects in the reading behaviour of LBL readers (especially in cases who are a priori diagnosed as having a peripheral impairment). Because the interest has traditionally been on the effect of word length, unconfounded by other variables, stimuli are standardly matched on frequency and imageability, for example, and length alone is manipulated. In those few cases where these other variables are tested and reported, performance is measured in terms of accuracy rather than the more sensitive measures of reaction time and the statistical analyses of both the main effects and their interaction are often omitted. Nevertheless, we have reviewed the literature and present the findings in Table 1. We discuss covert or tacit reading, the word superiority effect, and the influence of lexical variables separately.

Positive evidence for preserved implicit reading under brief exposure is found in only a small number of cases. Aside from the studies by Coslett and Saffran and colleagues, who document covert effects in five LBL readers, definitive covert processing is reported only in patients ML (Shallice & Saffran, 1986), KW (Howard, 1991) (in semantic but not lexical decision tasks), DR (Doctor et al., 1990), and RE (Feinberg, Duckes-Berke, Miner, & Roane, 1995). A positive trend in this direction is also noted in Grossi, Fragassi, Orsini, Falco, & Sepe (1984) and in Landis, Regard, and Serrat (1980), and we have some preliminary evidence of relatively good semantic categorisation under brief exposure in one of our own patients, DK. By contrast, the absence of covert abilities has been documented in several studies (Howard, 1991; Patterson & Kay, 1982; Price & Humphreys, 1992, 1995; Vigliocco et al., 1992; Warrington & Shallice, 1980) and in some of our own studies (Behrmann, Black, & Bub, 1990; Behrmann & McLeod, 1995; Behrmann & Shallice, 1995). Why there are such marked individual differences is taken up further in the General Discussion.

With respect to the word superiority effect, as with the covert reading, only a small subset of the population has been tested and so strong claims about the effect of orthographic context should be made cautiously. Of the 16 reported instances, 12 patients show a word superiority effect, one shows a trend in that direction, and the remaining 3 are not influenced by orthographic context. One of these 16 patients is DS who, in our initial testing in the first year post-stroke, did not show a word superiority effect (Behrmann et al., 1990) on a Wheeler-Reicher type task. In subsequent testing, roughly 5 years post-stroke, DS did show a word superiority effect on one task, reporting the beginning and end letters more accurately from words than from nonwords. DS, however, did not show an advantage for words in a second task that required a decision on whether two strings (both words or both nonwords) placed one above the other for an unlimited duration were the same or different (Behrmann & Shallice, 1995).

A suggestion for why the word superiority effect is only observed in some but not other cases has been made by Farah and Wallace (1992; see also Farah, 1997). They observed that at least two of the patients who showed the word superiority effect had a letter recognition profile that was similar to normal subjects (referred to as an "ends-in" profile). At least two of the cases who did not show a word superiority effect showed a gradient of letter recognition, with best performance on the initial letter and decreasing accuracy across serial position. They attribute the presence of a word superiority effect to the presence vs. absence of the letter-by-letter gradient. In fact, even within a single subject, one can see this at work: Bub, Black, and Howell (1989) found that their patient showed a word superiority effect in only one experiment and it was in this single experiment that the patient did not show a left–right gradient in accuracy across letter positions. The suggestion that strategy determines the presence of a word superiority effect may well explain the variance in the data on this effect in LBL readers. In fact, even in normal subjects, focusing sequentially on individual letters affects top-down effects; specifically, the word superiority effect is reduced when normal subjects attempt to read letters in particular positions rather than distributing their attention across the entire stimulus array (Johnston & McClelland, 1980). Unfortunately, the patients who show covert reading are generally not tested for a word superiority effect and vice versa, making it difficult to reach conclusions about the relationship between implicit reading and higher-order orthographic processing.

Of the 26 subjects tested for effects of frequency on word naming, 17 are influenced by frequency either in reaction time or in accuracy, including all 7 of our patients. A further six subjects show some effect of frequency (statistics are not always available) and only three subjects are not obviously influenced by frequency. One of the 3 subjects who does not show a frequency effect is patient RE (Feinberg et al., 1995), whose performance is so poor that he is unable to read even a single word aloud. The absence of a frequency effect then might simply result from a floor effect in this patient.

Of the 19 subjects tested for imageability in naming, again in reaction time or in accuracy, 12 show a positive result (5 of our 7 patients). Again, one of the subjects who does not is patient RE, and this may again result from a floor effect. Five out of 14 subjects show an effect of regularity on naming and 5 out of 9 subjects show an effect of part-of-speech on their performance. Few subjects are tested on these lexical variables in lexical decision, as is evident in the final two columns of Table 1. All eight patients for whom frequency data are available for lexical decision are affected by frequency. There are imageability data for only five patients in lexical decision, and two of these show a positive finding.

Taken together, the review of the later lexical and semantic effects is not strongly conclusive. Too few patients are tested on the higher-order variables and when they are, they are usually not tested on all of the different tasks or different variables, making comparisons difficult. One finding worth noting, however, is that across the population of 57 subjects, there are only 13 subjects who do not show any lexical or semantic effects on any of the measures. A further four patients have not been tested on any of the measures and so the presence of the later effects for them remains indeterminate (Lazar & Scarisbrick, 1973; Montant, Nazir, & Poncet, this issue; Stachowiak & Poeck, 1976; Warrington & Shallice, 1980; patient JDC). A central finding of the review, then, is that the great majority of patients show some form of lexical/semantic effects on reading performance.

Summary of Review

The evidence for a peripheral deficit is strongly supported by the review and there is no convincing counter-evidence. In those few cases where accuracy or even reaction times appear to be within normal limits, it is often the case that this could result from a speed–accuracy trade-off and poorer performance is generally seen on the complementary metric. These results substantiate the initial part of our account and endorse the previous observation of a deficit in letter processing that is common to all LBL readers. Unfortunately, strong conclusions about later lexical and semantic effects are not as obvious from this review, as there are too many empty cells in the database. An important finding, though, is that these later effects are not restricted to a small subgroup of the population and there is at least partial evidence for them in more than two-thirds of the patients.

EMPIRICAL DATA FROM LETTER-BY-LETTER READERS

Given the paucity of data on the lexical and semantic effects in LBL patients, we have undertaken a retrospective analysis of data collected from seven who have participated in our studies over the last decade. In this analysis, we specifically explore whether both a peripheral visual processing deficit and the effect of lexical and semantic variables are observable across this group. All of these patients are highly accurate in single-letter identification under unlimited exposure duration although, as pointed out previously, this in no way indicates that letter recognition is normal and, in some instances, we know that it is not (for example, patient DS; see Behrmann & Shallice, 1995). It is also the case, though, that these patients are much more impaired than their control counterparts in identifying two or three random letters presented under brief exposure (for example, patient DS, see Behrmann, Black, & Bub, 1990; patient IS, Behrmann & McLeod, 1995).

Our view of LBL reading predicts that, even in those patients with a peripheral deficit, one would still see effects of word frequency and imageability on performance. On this account, we therefore predict significant main effects of word frequency and imageability that will reflect the strength of the top-down contribution. We also predict that as word length increases, the difference between high- and low-frequency and the difference between high- and low-imageability items will increase, given that for longer words there is more opportunity for top-down effects to influence performance.

Subjects

Seven LBL readers are included in this analysis. Five of them have been described in detail in other publications (see final seven rows of Table 1 for citations), whereas the remaining two, PC and DK, have only been tested fairly recently. The subjects will only be described briefly in this paper and the reader is referred to the more detailed publications for further information. All subjects are right-handed and native speakers of English. Biographical information and anatomical details of their lesions as well as some reading data (which will become relevant later) are provided in Table 2.

Evidence for an Early, Peripheral Deficit

As mentioned previously, all of these patients do reasonably well, although not perfectly, on single-letter identification under unlimited exposure duration, although performance is markedly poorer than normal subjects when brief exposure is used. For example, patient DK performed perfectly on single-letter identification when the letters were presented for an unlimited duration. He made 30% errors when the letters were presented for 250msec (even without a mask) and 40 errors at 100msec exposure duration. Similar patterns are seen in other patients. Those patients who perform well on single-letter identification even at brief exposure duration may then do poorly when two or three letters are present; patient IS, for example, did well with single letters even at the presentation limits of the computer (17msec; no masking) but performance was poor on two- and three-item arrays even when exposure duration was more than tripled (Kinsbourne & Warrington, 1962; Levine & Calvinio, 1978). Relative to normal subjects, who can identify letters at 10msec with masking (Sperling & Melcher, 1978), these patients are much slower at letter identification.

Four of the seven subjects participated in the study by Sekuler and Behrmann (1996), which documented the ability of LBL patients to process visual stimuli that are not alphanumeric. In that study, patients completed three different experiments: they performed perceptual fluency judgements under time pressure, made same/different judgements to contour features that appeared on objects, and searched for a target in visual displays that increased in the number of distractor items. In all three experiments, these patients performed significantly more poorly than their control counterparts. These findings suggest that all four of these patients suffered from a general perceptual deficit, which extended beyond their ability to deal with orthographic material. It is this more fundamental visual perceptual problem, we claim, that gives rise to their LBL pattern in the first instance.

Five of these same seven subjects (excluding PC and TU) have also participated in a recent study examining their accuracy and reaction time to name black-and-white line drawings of high and low visual complexity, taken from the Snodgrass and Vanderwart (1980) set of pictures (Behrmann et al., 1998). Whereas age-

Table 2. Biographical and Lesion Details of the Seven Letter-by-letter Readers

Patient	Age[a]	Etiology	CT Scan Results	Other Relevant Behaviours	Freq[b]	Image[c]
DS	37	Posterior cerebral artery occlusion; migrainous	L occipital infarction	upper quadrantanopia	397**	72**
PC	63	Resected meningioma	L occipitotemporal (area 19/37 boundary)	R homonymous hemianopia	1651*	3039**
MW	67	Infarction	L occipital lobe infarction	R homonymous hemianopia; ensuing depression	570*	292**
DK	65	Posterior cerebral artery infarction & mass effect	L occipital lobe infarction	R homonymous hemianopia	588**	443
MS	37	Closed head injury	No focal CT lesion; bilateral frontal slowing EEG	R homonymous hemianopia; surface dysgraphia	1552**	1629*
IS	46	Posterior cerebral artery infarction	L occipital-temporal region including hippocampus, fusiform and lingual gyri	R upper quadrantanopia; mild memory deficit; surface dysgraphia	540**	99*
TU	56	Resected arteriovenous malformation	L occipital haemorrhage; additional L temporal damage	R homonymous hemianopia; R hemiparesis; marked anomia	968**	388

[a] Age refers to the age at which the initial testing took place. Some patients have participated in subsequent follow-up studies and the age of testing is then obviously different.
[b] Frequency score in msec.
[c] Imageability score in msec.
*$P < .05$; *$P < .01$.

matched normal control subjects showed a 152msec reaction time advantage for low-complexity items over high-complexity items, the LBL readers (with the possible exception of patient MW) showed a disproportionate increase in reaction times to name the high-complexity items. Averaged across the patients, low-complexity items were named 530msec faster than their high-complexity counterparts. Taken together, these data suggest that these patients have a deficit that affects their processing of several classes of visual stimuli and that, under rigorous testing conditions, this more widespread perceptual deficit may be uncovered.

Based on these findings, we can definitively conclude that there is an early deficit in six of the seven LBL readers. We do not have sufficient information about PC, but the limited data obtained from her single-letter processing suggest that she too may be impaired in early stages of processing prior to activating an orthographic representation. Whereas she is able to identify accurately all three letters from a random letter string (e.g. YFS) presented for an unlimited exposure duration, she only identifies about 70% of the letters when the exposure duration is around 1sec, suggesting that her letter identification threshold is probably abnormally high. Unfortunately, PC died before we were able to obtain any further data.

Evidence for Lexical and Semantic Influences on Word Recognition

There are a number of different lexical variables that might potentially serve as markers of higher-order processing in a naming latency task, including effects of word frequency, differences between word and nonword reading (word superiority), part-of-speech effects, and an effect of imageability or concreteness (Coltheart, 1980; Warrington & Shallice, 1980). We have chosen to use word frequency and imageability, both of which are assumed to be reliable indicators of access to a preserved word form system (Kay & Hanley, 1991), and for which there is likely to be sufficient data for analysis. These two variables have also been studied extensively in normal readers and there are several models detailing their specific effects in word recognition. A recent analysis of the effects of frequency, for example, shows that word frequency is a critical determinant of the speed of lexical access (what Monsell et al. referred to as "lexical identification") (Monsell, Doyle, & Haggard, 1989). Frequency effects can be explained, depending on whether one's model is localist or distributed, as arising from higher resting levels of activation, lower thresholds, or increased bias on the weights for high-frequency compared with low-frequency words (McClelland & Rumelhart, 1981; Morton, 1979; Plaut et al., 1996; Seidenberg & McClelland, 1989). Thus, the speed by which a word is recognised is determined by the time to reach threshold or for the network to settle on an interpretation of the stimulus; the time will thus be less for high-frequency than for low-frequency items.

Imageability effects have not received as much press as frequency effects, although there is little doubt that the influence of imageability on performance arises at later stages and is indicative of semantic processing. Strain et al. (Strain, Patterson, & Seidenberg, 1995), for example, showed that normal readers are slower and make more errors on low-imageability than high-imageability words but only for low-frequency exception words. This three-way interaction between imageability, frequency, and regularity is explained with reference to a system in which the time to translate orthography to phonology varies. When this translation process is slow or noisy, as is the case for low-frequency exceptions, words with rich semantic representations (i.e. high-imageability words) are most likely to benefit from this interaction. This interpretation of how later effects (semantic representations in this case) interact with somewhat earlier processes (translating orthography to phonology in this case) is consistent with the interactive account of LBL reading that we have proposed.

Stimuli and Procedure

To examine the effects of frequency and imageability on the reading behaviour of LBL readers, we analysed naming latency data for single words collected from these seven patients. At some point during their testing, these patients read aloud lists of 60 words presented in a variety of fonts and sizes containing 20 items each of 3-, 5-, and 7-letter words. These lists and forms of presentation are used standardly in our laboratory to measure the naming latency of LBL readers and patients typically read three of four such lists during their preliminary testing. We have included as many word lists as possible for each subject apart from those lists in which the words were presented in script or cursive font. Frequency and imageability are orthogonally crossed with word length in each list. The cut-off for frequency was 20 per million, with items below that classified as low in frequency and items above that classified as high in frequency (Kučera & Francis, 1967). Imageability is standardly hand-coded but, for the purpose of this analysis, word imageability was taken from the MRC Database; items which did not have an imageability rating in this listing were excluded. Again, imageability was converted to a categorical variable with items exceeding 525 being classified as high in imageability and those below this cut-off as being low in imageability.

In all cases, the words were presented on a computer screen to the left of central fixation to circumvent the right visual field defect that was present in all patients. The words were right-justified so that the final letter of each word appeared in the character space just to the left of fixation. The standard procedure was as follows: a central fixation point appeared for 500msec. After a 1 sec delay, the word appeared and remained on the screen for an unlimited duration until a response was made. The visual angle subtended by stimuli of three, five- and seven-characters in length were 1.5°, 2.4° and 3.6° respectively. Words were presented in black font against a white background. Both reaction time (via a voice-relay-key) and accuracy (recorded by the examiner) was measured. In all patients, except in the case of DK, presentation and timing was controlled by Psychlab (Bub & Gum, 1988) and a Macintosh Classic or IIci was used, whereas for DK, presentation was controlled by PsyScope (Cohen, MacWhinney, Flatt, & Provost, 1993) on a Powerbook 540C.

Results

Reading responses to a total of 24 word lists ($N = 1440$, cumulative across subjects) were collected from the subjects. A number of trials, however, were excluded for a variety of reasons: imageability ratings were unavailable for 315 trials, patients made errors on 168 trials, the microphone was not triggered or the subject coughed on a further 50 trials and 2 trials were extreme outliers from the subjects' reaction time distribution (more than four SDs from the mean). A repeated-measures ANOVA with word length crossed with word frequency and word imageability was performed on the remaining 905 trials. Subject was included as a further between-subject variable so that profiles for the individual subjects could be drawn up. We note that the amount of data per subject is not equivalent (given that different subjects read differing numbers of lists and error rates differed) and so the estimates of the effects are better for some subjects than for others; subject PC has the least data and her performance is the most variable (see all Figs). As mentioned earlier, unfortunately PC died before we were able to collect further data. The number of trials per subject is included in both Figs 2 and 3.

The reaction time (RT) to name a word as a function of string length is plotted for each subject individually in Fig. 1. Along with each line is the value of the slope calculated for that subject in a linear regression analysis, with RT set against word length. As is evident from this figure, subjects differ fairly dramatically in their base reaction time [$F(6,821) = 118.7$, $P < .0001$]. Also evident from this figure is the significant main effect of word length [$F(2,821) = 167.3$, $P < .0001$], with a mean across all subjects of 1442, 2465, and 3520msec for 3-, 5-, and 7-letter words, respectively, indicating that roughly an extra 500msec is required to process each additional letter.

Although every subject shows the hallmark monotonic relationship between reaction time and word length, subjects differ in the slope or increase in RT per-each additional letter [$F(12,821) = 18.3$, $P < .0001$]. At the highest extreme is MA, who shows an increase of 1409.3msec per letter. At the bottom extreme, the mildest LBL subject, DS, who had a stroke roughly 10 years ago (Behrmann, Black, et al. 1990), shows an increase of 97msec per letter. Although this latter slope is rather flat in comparison with the population of LBL readers, it must be noted that this increase is roughly three times greater than the maximum of 30msec per letter needed by normal subjects in naming words presented to the left visual field (Henderson, 1982; Young & Ellis, 1985). Indeed, some studies have demonstrated that normal subjects show no effect at all of word length on naming latency for words (Schiepers, 1980; Weekes, 1997). Any reliable increase in latency as a function of string length might, therefore, be considered abnormal. On the same word lists as those read by our LBL readers, normal subjects, who served as control

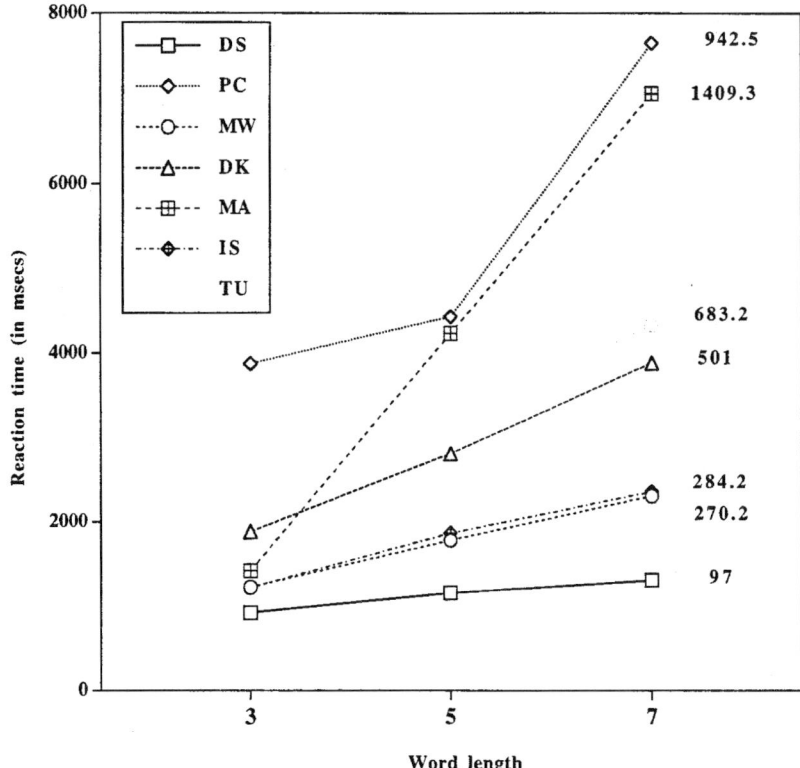

Fig. 1. Mean reaction times for the seven letter-by-letter readers as a function of word length and the slope of the regression function.

subjects for some of the patients (some of whom are elderly), typically showed very flat slopes; control subject BR (Behrmann & McLeod, 1995), for example, showed an increase in RT of 15.3msec for each additional letter in a word, whereas control subjects AS and JD, averaged, show a slope of 9.4msec (Behrmann et al., 1998). Relative to these values, all of our subjects are markedly abnormal, including the mildest subject, DS.

Importantly for the present purposes is that, across these seven subjects, word frequency significantly affected reaction time [$F(1,821) = 41.8$, $P < .0001$]; high-frequency words ($N = 566$) were named, on average, 884msec faster than low-frequency ($N = 339$) words. The differences between the mean RTs for high- and low-frequency words for each patient are included in Table 2. A one-way ANOVA performed separately for each subject, comparing RTs for high- and low-frequency words, yielded significant values for every subject. The effect of frequency did not vary as a function of word length, collapsed across subjects [$F(2,821) = 2.06$, $P < .13$], and the frequency effects remained relatively constant across subjects [$F(6,821) = 1.17$, $P < .32$]. However, the three-way interaction of frequency by word length by subject is significant. This is seen in Fig. 2 which shows the mean reaction times for the individual patients for high- and low-frequency words as a function of word length. It is important to note that the y-axis on this figure differs across subjects. As is evident from this figure, the difference between high- and low-frequency words increased as a function of word length, although this held to a varying degree across subjects [$F(12,821) = 1.76$, $P < .05$]. The interaction is particularly evident in patients DS, MW, DK, and IS and, to a lesser degree, in MA. It is only in patient PC (for whom the least data are available) that this function does not follow the predicted pattern. Interestingly, there is a good correlation across the subjects between the severity of the LBL reading (defined by the slope of the regression curve on the naming latency data in Fig. 1) and

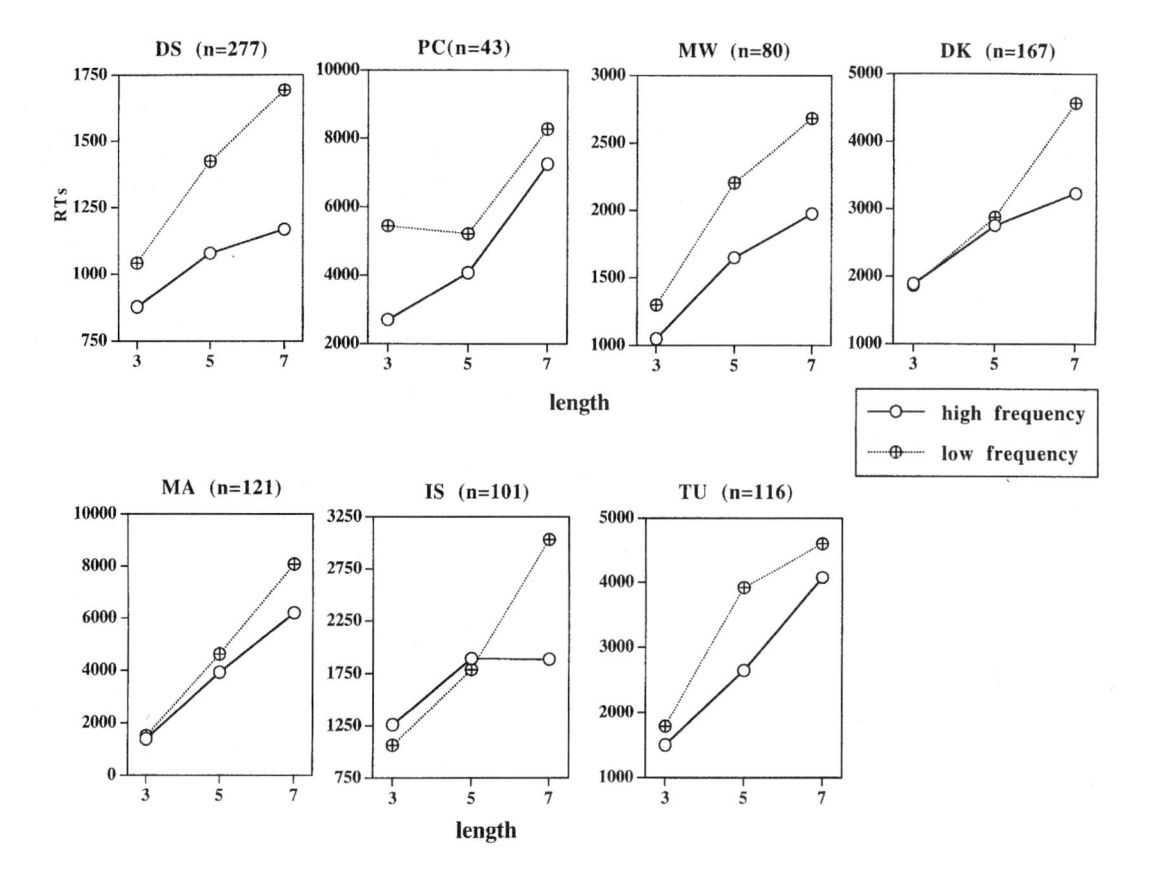

Fig. 2. Reaction times for the seven letter-by-letter readers, plotted individually, as a function of word length for high- and low-frequency words.

the effect of frequency on performance (defined as the msec difference between naming high- and low-frequency words ($r^2 = .85$, $P < .005$), indicating a close relationship between severity and frequency, as would be predicted by our account.

Word imageability also significantly influenced naming [$F(1,821) = 52.3$, $P < .0001$]; high-imageability words ($N = 455$) were reported 696msec faster than low-imageability words ($N = 450$) on average, although the extent of this difference varied across subjects [$F(1,821) = 10.6$, $P < .0001$]. The differences in mean RT for high- and low-imageability words for each subject are presented in Table 2 along with the significance values obtained from a one-way ANOVA performed separately for these data. Five of the seven patients showed a significant difference between high- and low-imageability stimuli. The advantage for high- over low-imageability words was, however, influenced by word length [$F(2,821) = 20.7$, $P < .0001$]; the difference between high- and low-imageability words is 169msec for 3-letter words compared with an 852msec difference for 7-letter words. Again, this difference as a function of word length varies across subjects. This is evident in Fig. 3 in which the mean reaction times for the individual patients with high- and low-imageability words are plotted as a function of word length. Although there is a general trend in all subjects (except for subject DS) for the difference between high- and low-imageability words to become increasingly magnified as word length increases [$F(12,821) = 7.6$, $P < .0001$], these data are more variable than the frequency data and the pattern across subjects is less clear. The expected interaction, however, is clearly evident for patients PC, MW, DK, MA, and TU and less clear in DS and IS (although the increase in 5-letter words is perhaps a bit more evident for IS). It is also worth noting that PC's data are particularly variable; PC has a total of only 43 data points, 5 of which are 7-letters long and low-imageability, and thus the very long mean RT of 13049msec for this last cell should be interpreted cautiously. There was a positive correlation between the severity of the LBL deficit (again, using the value of the regression slope in naming latency) and the imageability effect (defined as the msec difference between naming high- and low-imageability words) ($r^2 = .53$, $P < .06$), again indicating a relationship between severity and imageability, although this correlation was not as strong as was the correlation with frequency.

Summary of Empirical Data

The findings from the empirical study are fairly straightforward. In a group of LBL patients, all of whom show the characteristic effect of word length on reaction time, we see a difference in performance between high- and low-frequency items in all seven patients and a difference between high- and low-imageability words in five of them. Moreover, both frequency and imageability interact with the length of the string such that the frequency effect and the imageability effect are more marked for long than for short words. The exact discrepancy between the high and low items as a function of word length is evident to

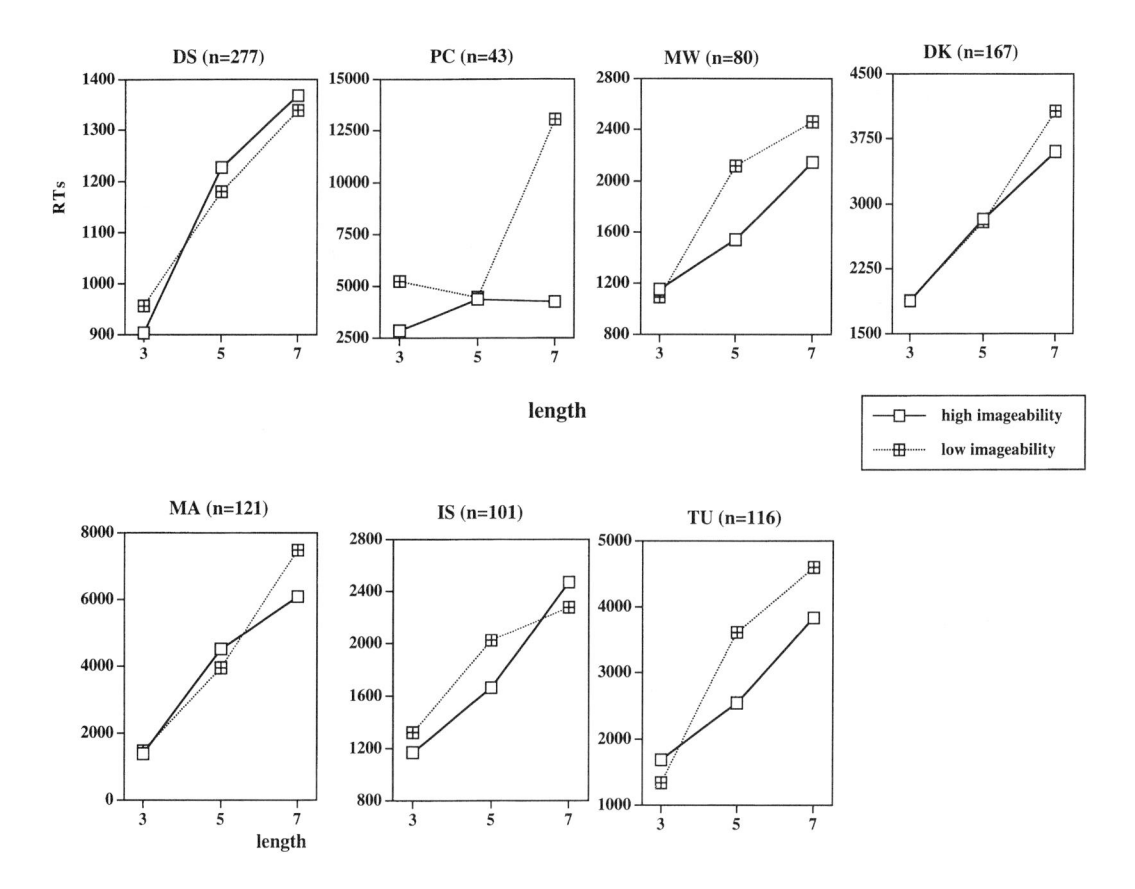

Fig. 3. Reaction times for the seven letter-by-letter readers, plotted indivdually, as a function of word length for high- and low-imageability words.

a greater degree in some patients than in others and this interaction is more pronounced in the frequency than in the imageability analysis.

DISCUSSION

Letter-by-letter reading is a form of acquired dyslexia in which patients show a monotonic relationship between word length and reaction time, as reflected in both naming and lexical decision tasks. There are two major findings in the LBL literature: The first is that patients have a deficit at a peripheral stage of processing. Evidence supporting a peripheral impairment comes from the finding that many, if not all, patients are unable to represent orthographic input normally and they show an impairment in letter processing. The second finding is that the reading performance of many patients is influenced by lexical and semantic variables, suggesting that the deficit occurs only after an adequate orthographic representation is activated. Support for a central impairment comes from the finding that many LBL readers appear to have processed the visual stimulus sufficiently to have derived semantic, lexical, and higher-order orthographic information from it, as indicated by their ability to perform semantic categorisation tasks or to make lexical decisions about stimuli that they have not identified explicitly.

In this paper, we put forward an account of LBL reading that reconciles these disparate findings. We argue that there is a fundamental peripheral deficit in LBL that adversely affects parallel letter processing. Despite this lower-level deficit, however, some information is still propagated to higher-level (lexical and semantic) representations. In an attempt to enhance orthographic activations, subjects employ the normal reading strategy of making additional fixations (or covert attention shifts) when encountering difficulties in text. This sequential letter processing adds further to the activation of lexical/semantic representations. Due to the interactive nature of processing, these higher-level representations feed back to provide support for subsequent orthographic processing. The strength of this support depends on the degree of higher-level activation, which we assume scales with word frequency and imageability. Because patients take longer to process words with more letters, the lexical/semantic activation has a longer time in which to accumulate and influence the degraded letter activations. The main point is that, in an interactive system, impoverished input can still activate lexical and semantic information and, moreover, that the strength of this activation over time is related to the length of the word. Thus, covert or tacit reading in the form of semantic categorisation and lexical decision is possible despite poor stimulus identification; lexical variables such as frequency and imageability may still influence word recognition, and lexical/semantic representations may feed back onto letter representations and produce a word superiority effect. Thus, just as higher-level representations can compensate for or mitigate against the spatially degraded input in neglect dyslexia (Mozer & Behrmann, 1990), so too can such representations feed back and support the weakened orthographic activation in LBL reading.

In support of this theoretical position, we cite evidence from a review of published reports of 57 patients with LBL reading. There was positive evidence in almost every case for the presence of a peripheral deficit and, even in those few cases for which no definitive evidence existed, there was no compelling counter-evidence. The findings from this review regarding the higher-level lexical and semantic effects were somewhat less conclusive but this was largely because of the dearth of data. Relatively few patients have been tested systematically for the influence of different variables on performance and, even when there was some suggestive evidence in a case report, the effects were not usually evaluated across a sufficient range of measures to provide a full description of the patient's performance. Perhaps the most relevant finding from this review, however, was that more than two thirds of the patients showed some evidence of higher-level effects, whether in implicit reading under brief exposure, in the presence of a word superiority effect, or in the influence of lexical variables on reading performance. The key point, then, is that in many of these patients with peripheral deficits, there is also a strong suggestion of some kind of lexical and semantic processing.

To substantiate our position further, we carried out a retrospective empirical analysis of the reading performance of seven LBL readers we have tested over the last decade. We have previously determined that these patients all suffer from a peripheral impairment of some form. In the current analysis, we demonstrated a significant main effect of word frequency and word imageability on reaction time in almost all cases and, in most cases, these effects interacted significantly with word length (although to varying degrees and with more variance in imageability than in frequency). We suggest that this interaction is specifically predicted by an interactive account. In particular, we argue that, because of the additional time required by LBL readers for longer words, there is more time in which higher-level activation can accumulate and affect reading performance.

Our review of the literature has turned up a number of other cases in whom an interaction between a lexical variable and word length has also been demonstrated. For example, Bowers, Bub, & Arguin (1996) documented the effect of word frequency on the accuracy and reaction time of patient IH's reading responses. Although a statistical analysis was not provided, high-frequency words were read better and faster overall than were low-frequency words.

Moreover, the discrepancy between the two types of items was greater for longer than shorter words: whereas there was no accuracy difference for high- and low-frequency four-letter words, there was a difference for longer words. Interactions of word length and imageability have also been reported in a few cases. Patient DR (Doctor et al., 1990), for example, took longer to read low- than high-imageability printed words, particularly as word length increased. A similar trend, albeit nonsignificant, was also observed in this patient for handwritten words. Finally, JH (Buxbaum & Coslett, 1996) showed a dramatic increase in reaction time (and decrease in accuracy) as word length increased, but this was especially true for low-imageability words. Thus, the difference in accuracy between high- and low-imageability words was approximately 10% but increased to approximately 45% for eight-letter words (see their Fig. 7).

In addition to showing the predicted interaction between word length and imageability, patient JH (Buxbaum & Coslett, 1996) provides additional support for our account. JH was clearly impaired at lower-level processing: He was unable both to represent information on the right of a display and to distribute attention with sufficient resolution for the purposes of letter identification. Despite this impaired peripheral processing, however, JH's reading performance showed a part-of-speech effect (nouns were read better than functors), an effect of lexical status (words were read better than nonwords), and an effect of morphological composition (pseudosuffixed words were read better than suffixed words). The two major findings, a peripheral deficit concurrent with lexical influences on reading, and an interaction between word length and lexical variables, provide clear support for our unitary view of letter-by-letter reading.

There are a number of suggestions in the literature that bear some resemblance to, or even foreshadow, the view we have proposed here. More than a decade ago, Shallice and Saffran (1986), in their explanation of patient ML's tacit or covert reading abilities, suggested that weak input from an impaired word-form system might still allow sufficient activation of a semantic representation that would suffice for a task like semantic categorisation. They argued, however, that inhibition might be inadequate in the system. Because it is necessary to converge on a single correct output for stimulus identification, the alternatives must be inhibited and patient ML was unable to do so. Thus, weakened but sufficient activation in combination with inadequate inhibition might give rise to above-chance semantic categorisation in the absence of stimulus identification (see Friedman et al., 1993, for a similar account and Hinton & Shallice, 1991, for an interpretation of ML's data that is consistent with this). Along similar lines, Bub et al. (1989) have suggested that, if activation is sufficient, even if not of normal strength, lexical decision and semantic priming may still be possible. Similar ideas about weakened activation and/or inhibition have also been suggested to explain phenomena of tacit recognition in other neuropsychological disorders, such as prosopagnosia or implicit/explicit memory differences (Farah, 1994), and even in normal subjects. Monsell et al. (1989), for example, have postulated a mechanism similar to ours to account for the finding that normal subjects can perform lexical decision well without being able to identify the stimulus. They claimed that a subthreshold signal from the target relative to the activity of neighbouring items may enable a legitimate word to be distinguished from nonwords but may not suffice for identification. If the subjects monitor the activation in the system, "the familiarity assessment process", then high-frequency words which have stronger representation and are more familiar can bias the subject towards a positive response before unique identification has occurred (see Plaut & Shallice, 1993, for connectionist simulations consistent with this idea).

MODELS OF LETTER-BY-LETTER READING

There are no existing models of reading that fully instantiate our theory of the mechanism underlying the performance of LBL readers. However, there are a number of models which, in their behaviour under damage, exhibit many of the critical properties on which our account is based. As mentioned in the Introduction, the fact that a partial peripheral impairment may give rise to both lower- and higher-level effects has already been demonstrated with respect to lexical influences on reading performance in neglect dyslexia (Mozer & Behrmann, 1990), in the preservation of semantic categorisation and lexical decision with impaired overt recognition, and in the occurrence of semantic errors following visual damage in deep dyslexia (Hinton & Shallice, 1991; Plaut & Shallice, 1993). In the specific context of LBL reading, support for co-occurrence of lower- and higher-level effects following peripheral damage has recently been provided by Mayall and Humphreys (Mayall & Humphreys, 1996) who implemented a connectionist model whose architecture incorporated some of the properties of the IAM. In a single architecture, reflecting the normal word processing system, Mayall and Humphreys were able to reproduce many of the covert abilities of LBL patients when they lesioned the network. They tested both early lesions (near the input layer) and more central lesions (to hidden units). Of primary concern to us is the pattern of results obtained with the former, peripheral type of disturbance. When a percentage of the input units were disconnected from the hidden units, letter naming, although poor (35% correct) was significantly better than word naming (9% correct). The errors produced were mostly visual rather than semantic. In addition, a greater number of correct semantic categorisation and lexical decision responses were observed compared with naming responses. Finally, effects of word frequency and imageability were noted in naming, categorisation, and lexical decision. This ensemble of findings is precisely what one would expect with a cascaded, interactive model.

A significant limitation of Mayall and Humpreys' (1996) simulation, however, is that it does not address the hallmark characteristic of LBL readers—namely, their letter-by-letter reading. More recently, Plaut (1998) has demonstrated properties of LBL reading following peripheral damage to the "refixation" model first described by Plaut et al. (1995). The model generates a sequence of phonemes as output in response to orthographic input presented over position-specific letter units. A critical aspect of the model is that, if it encounters difficulty in generating a particular phoneme in the course of pronouncing a word, it can refixate the input (using an internally generated attentional signal) to bring the corresponding peripheral orthographic segment to "fixation", where performance is better. Early on in training, the model fixated virtually every grapheme but, by the end of training, it read correctly 99.3% of the 2998 monosyllabic words on which it was trained, producing an average of only 1.3 fixations per word. When letter activations were corrupted by noise, correct performance dropped to 90.1% correct. Using a median split on frequency, accuracy was greater on high- versus low-frequency words (92.1% vs. 88.6%, respectively) and short versus long words (e.g. 92.0% for 4-letter words vs. 85.2% for 6-letter words). Critically, among words pronounced correctly, the average number of fixations per word increased from 1.3 to 1.93. Moreover, the number of fixations—a loose analogue to naming latency—was strongly influenced by the length of the word. For example, the model made an average of 1.76 fixations for 4-letter words but 2.92 fixations for 6-letters words. The model also made fewer fixations to high- versus low-frequency words (means 1.86 vs. 2.00, respectively). Finally, and most importantly for the current account of LBL reading, there was a clear interaction of frequency and length: The difference in the number of fixations for high- versus low-frequency words was 0.04 for 4-letter words but 0.20 for 6-letter words. Thus, under peripheral damage, the model exhibited the hallmark word-length effect characteristic of LBL reading, combined with the appropriate higher-level effects: A word frequency effect that was greater for long compared with short words. However, the model contains no semantic representations and, thus, is unable to account for the effects of imageability on LBL reading, nor the relatively preserved lexical decision and semantic categorisation performance of these patients.

ADDITIONAL ISSUES

There are three outstanding issues that require discussion in the context of our theory of LBL reading. The first concerns effects of stimulus degradation in normal readers, the second concerns the heterogeneity in the performance of LBL readers, and the last concerns the substrate mediating the lexical and semantic effects. We take up these issues in turn.

Effects of Stimulus Degradation in Normal Readers

A straightforward prediction that one might make from our account is that normal subjects should reveal a word length effect when tested under conditions of stimulus degradation (and that this should interact with frequency and imageability). Exactly what the parameters of this degradation are remains unclear and needs to be determined. There are, however, a few studies that have examined this issue in normal subjects. An early study by Terry, Samuels and LaBerge (1976; Expt 1) had normal subjects press a response button when the presented item was an animal word (two thirds of all words were animal words and were the highest-frequency animal words). The words varied in length from three to eight letters and the letters could be degraded or not. Unfortunately, conclusions from this study are tenuous for our purposes. First, it appears that the degradation had no effect on accuracy, as evident in the error data (op. cit. p. 580). Given that this is a potential ceiling effect, it is not surprising that there is no interaction between degradation and word length even in reaction times. Second, the results for the longer words (seven and eight letters) could not be calculated as there were too few observations in these cells. Finally, because the task is more like a decision task than a naming latency task, the effect of word length, if it exists, is likely to be smaller than is the case in standard naming latency tasks (see Henderson, 1982, for review of relevant findings).

In a more recent study, Snodgrass and Mintzer (1993), have examined the effect of stimulus degradation on word length in the context of studying how orthographic neighbourhood and word frequency influence the speed of word recognition. They presented degraded word stimuli of varying lengths to normal readers and gradually improved the quality of the stimulus until it could be identified. The threshold of time to identification did not vary as a function of word length nor did it vary as a function of neighbourhood or frequency, as one might have expected (see Snodgrass & Mintzer, 1993; Table1, p. 251). The absence of a correlation between the threshold of identification and any of these other variables, as acknowledged by the authors, might be a function of the fact that frequency and neighbourhood size were not particularly carefully controlled. In light of this, the authors go on to a second experiment in which these variables are controlled and the expected results obtained. Unfortunately, length is not controlled in this second experiment and so conclusions about the effect of degradation of word length in normal readers remains to be determined.

A final experiment, perhaps the most closely related to the focus of this paper, is that of Farah and Wallace (1991) who plotted naming latency as a function of word length for stimuli that were either masked (as a proxy for stimulus degradation) or not. Whereas latency differed significantly as a function of word length and also of visual quality, there was no interaction between them. On the surface, this seems to be a challenge to our account. Important to note, however, is that Farah and Wallace (1991) instructed their subjects to read letter-by-letter as they were particularly interested in the visual impairment hypothesis of pure alexia. Unfortunately, for our sake, this does not constitute a relevant test of our account, given that strategic effects can significantly alter the nature of the reading processing (see following section on individual differences) and also that under these conditions, subjects are unlikely to make an initial attempt at parallel identification, as we propose is the case for LBL readers. Taken together, the findings from these three studies fail to show the expected interaction of word length and stimulus degradation. However, none of these studies appears to be a good test of the hypothesis given the particular methodological choices the experimenters have made. Our prediction still holds then: If one can simulate the nature of the peripheral impairment in LBL readers in normal subjects, then interactions with word length as well as the other lexical variables should be observed. Preliminary data are encouraging in that an interaction between word length with frequency and with imageability is more evident with increased stimulus degradation (Nelson, Behrmann, & Plaut, 1998).

Individual Differences in Letter-by-letter Reading

A particularly striking finding that seems to challenge to our account concerns the range of individual differences observed in the LBL readers. As is evident from Table 1, whereas some patients show striking lexical and semantic effects, others do not. There are several possible explanations for this variability. One explanation might be that this simply reflects limitations in the available data. Patients are not routinely and systematically tested and so the sparse data may be exaggerating the differences between patients. Although it is true that additional data are sorely needed, this is unlikely to be the sole explanation. It is also unlikely that individual differences in literacy or reading skill can fully explain the range of variability in the LBL reading profiles (Hanley & Kay, 1996).

Perhaps a more persuasive explanation for differences among LBL readers concerns the severity of the deficit. In Mozer and Behrmann's (1990) interactive theory of neglect dyslexia, the difference between the two patients is explained by the severity of the early attentional deficit. Thus, Mozer and Behrmann suggested that, in the case of the more severe neglect patient, the orthographic input was far too degraded to activate the corresponding higher-order lexical and semantic representations. In the case of the more mild neglect patient, there was sufficient bottom-up activation of higher-order representations, and effects of lexical status and morphological composition were observed. We suggest that differences in severity of impairments may also underlie the individual differences observed among LBL readers. For example, Behrmann, Black, et al. (1990) showed that, in the early stages post-stroke, DS did not show a word superiority effect in her reading. When she was retested several years later, her reading had improved (as measured in the slope and intercept of the latency-per-letter function), although she was still an LBL reader. Importantly, at this later stage, she now showed a marked difference in reading of words versus nonwords. Consistent with this, Coslett and Saffran (1989a) found that the postlexical effects in their patients became more apparent after partial recovery had taken place. For example, patient JG gradually regained the ability to perform tacit reading over a 3-month period and showed imageability and part-of-speech effects in reading (Coslett & Monsul, 1994).

These findings suggest that there may be an inverted-U function relating the severity of impairment to the strength of higher-level effects observed in LBL readers. When the visual input is well processed, top-down support is largely unnecessary, and when the deficit is too severe, higher-order representations are not strongly engaged. It is only in the middle range, when the orthographic input is still sufficiently intact, that the later lexical effects become apparent. At this point, then, when there is sufficient activation, the exact magnitude of the higher-order effects depends on the severity of the deficit. The slower the reader (or the more marked the word length effect), the longer the time for the top-down processes to play a role. This is well illustrated in our group of seven patients in whom there is a correlation between the severity of LBL reading (defined by the slope of the regression curve on the naming latency data; Fig. 1) and the frequency effect (difference in ms between high- and low-frequency words). There is also a positive correlation between slope values and the imageability effect, although this correlation is more modest than the frequency effects. Thus, the more severely affected the LBL reader (within the range permitting higher-level activation), the greater the effect of the lexical variables.

The claim that severity, defined this way, might determine the extent of the lexical/semantic effects, is interesting but is likely to be too simple. The slope of the reading function, in our account, is determined by the severity of the lower-level deficit, but we know that the correlation between the degree of peripheral impairment and reading performance is not perfect. Hanley and Kay (1996) have described fairly large differences in the reading performance of two LBL patients with roughly equivalent letter identification performance. Their interpretation of the discrepancy between the patients echoes that of Patterson and Kay (1982), who invoked a second lesion (in addition to the peripheral one) to account for the differences between patients. Specifically, they proposed that an additional impairment at the level of the word form system is present in some but not other LBL readers.

Some authors have suggested that yet another potential source of individual differences is parametric variation in the strength of overall inhibition within the lexical system. As mentioned previously, according to Shallice and Saffran (1986) it is the strength of inhibition that is critical for the system to converge on a single response for identification. Sufficient (even if not normal) activation, on the other hand, might be adequate for semantic categorisation. One might imagine that differing degrees of inhibition and activation might lead to different patterns of performance, only some of which produce results consistent with higher-level effects on performance. Arguin and Bub (1993), in their demonstration of letter priming in LBL reading in a connectionist model, have made a similar point. They argued that the crucial parameters are the amounts of inhibition and activation and the balance between them. It is this balance that may vary from one patient to another. Differences in response threshold and confidence in responding might also bring about differences in the overall pattern of reading performance across patients.

A final but important difference concerns the strategies individual subjects might employ in compensating for their peripheral impairment. It is now well recognised that particular strategies can diminish and possibly even eliminate lexical/semantic effects (also, Howard, 1991). Farah and Wallace (1991) pointed this out in relation to the presence/absence of the word superiority effect and the type of serial letter strategy used, whether "ends-in" or left-to-right. But perhaps the best illustration of the importance of considering strategy effects comes from patient JWC (Coslett et al., 1993), who appeared to be able to use two distinct strategies, only one of which produced covert reading. When JWC was instructed to name words, he employed a laborious, serial LBL strategy that eventually resulted in explicit word identification. In contrast, when instructed to make lexical decisions on semantic categorisations on briefly presented stimuli, he adopted a "whole-word" strategy that was fast and less effortful but which failed to provide explicit word identification. Based on these findings, Coslett and colleagues argued that differences in task demands or instructions may be a critical factor that affects the emergence of the higher-level effects and, presumably, in their account, the involvement of the right hemisphere in reading. They recommended that patients be discouraged from employing the slow and inefficient left-hemisphere LBL procedure during rehabilitation and should instead be prompted to adopt the more parallel whole-word strategy (but see Chialant & Caramazza, this issue.

The Locus of the Lexical/Semantic Effects

The last remaining topic for discussion concerns the debate between the right-hemisphere view and our own. The former, two-system account argues that the LBL reading and the word length effect are a consequence of the processing of the impaired left hemisphere, whereas covert reading and later lexical/semantic effects are mediated by the right hemisphere (Coslett & Saffran, 1993; Saffran & Coslett, this issue). We have maintained instead that the empirical data and profiles of performance associated with LBL readers do not require such a dichotomy. Rather, we have argued that the full range of effects in LBL reading can be explained in terms of properties of the normal interactive lexical processing system located in both hemispheres. Although the right hemisphere undoubtedly contributes to performance, as does the left hemisphere, the observed LBL pattern is a reflection of the combined processing of both hemispheres, both for the sequential reading pattern and the higher-level effects. After damage to the left hemisphere, we would argue, the dynamic co-operation between the hemispheres and their relative contribution to the output continues as was the case premorbidly. Importantly, on our account, the two hemispheres are governed by the same computational and interactive principles and the final manifestation of LBL reading is the result of the dynamic processing within and between both hemispheres.

What data, then, might compel us to abandon this view and invoke a stronger right-hemisphere account, or a greater division of labour between the hemispheres? One source of evidence often cited in support of the right-hemisphere account is the apparent similarity in the pattern of performance in LBL patients and in the right hemisphere of commissurotomy and left-hemispherectomy patients. These last two groups of patients can make semantic judgements but are poor at processing low-imageability words, grammatical morphemes, and functors—precisely the pattern reported for LBL readers (Coslett & Saffran, 1994). The qualitative similarity between the patient groups suggests that the LBL readers might well be accessing a limited-capability right-hemisphere reading system. There are, however, some important differences between these various populations both qualitatively and quantitatively (Baynes, 1990). For one, unlike deep dyslexic patients, who are thought to be reading primarily with the right hemisphere given the large extent of their left-hemisphere damage, LBL readers almost never make semantic paralexias. A possible response to this discrepancy is to claim that LBL readers are able to access the phonology for the initial letter via the left hemisphere and this prevents the production of semantic errors (a similar view is cited by Marshall & Newcombe, 1973, in the case of deep dyslexic patients who make fewer semantic paralexias over time). Even if this were the case, this would amount to the same claim as we are making: Both hemispheres contribute to the final output. Setting the semantic errors aside, however, if it were the case that LBL readers were engaging the right hemisphere to such an extent, we might expect that they would read words as well as patients with deep dyslexia. Some deep dyslexia patients read concrete words extremely well, and so the expectation is that this should also be the case for LBL readers. This, however, is not the case.

Perhaps the evidence to distinguish between these accounts will only come from physiological studies that directly investigate the contribution of the two hemispheres to reading in letter-by-letter patients. One such study is that by Coslett and Monsul (1994), which used transcranial magnetic stimulation (TMS) applied separately to the left and right hemisphere of their LBL patient, JG. Interestingly, JG's reading was significantly disrupted by TMS to the right (from 71% to 21%, and showed an interaction with frequency), but not to the left hemisphere (68% to 61%). This finding is strongly indicative of the contribution of the right hemisphere to reading. Interestingly, JG's reading was not totally disrupted by right-hemisphere TMS; this might indicate that either the TMS only partially disrupted the right-hemisphere function or it totally disrupted right-hemisphere function and the residual reading was mediated by the left hemisphere. Although this finding does seem, on the face of it, to suggest strongly that it is the right-hemisphere that is mediating reading, it should be pointed out that JG is a somewhat anomalous LBL patient. He exhibited pure alexia after suffering small, exclusively subcortical lesions to the splenium of the corpus callosum and in the region of the left lateral geniculate nucleus (he also showed a dilatation of the left occipital lobe on subsequent CT scans, perhaps reflecting degeneration secondary to the lateral geniculate damage). The subcortical nature of the damage is different from that in the overwhelming majority of LBL patients, whose lesions are cortical (Black & Behrmann, 1994). Thus, although the findings from JG are indeed provocative, we do not think that they are sufficiently firm on their own just yet to motivate a strong right-hemisphere account.

Our conclusion, then, is that there are no data that compel us to accept the view that the right hemisphere solely or even primarily subserves the lexical and semantic effects in LBL reading. Given this, we maintain that the normal premorbid reading system, degraded by brain damage, continues to function, and, because of its interactive nature, gives rise to the expected patterns of lower- and higher-level effects. Perhaps the final adjudication will come from neuroimaging studies of LBL readers in which brain activation of each hemisphere is calculated separately and correlated with the lexical/semantic effects.

CONCLUSION

We have presented a unitary account of LBL reading that, we have argued, reconciles previously discrepant findings. We have claimed that a deficit in letter processing (perhaps attributable to an even more fundamental perceptual impairment) is common to all LBL readers, and that this deficit prevents the normal activation of orthographic representations. It is the severity of this impairment that determines the extent to which top-down activation from higher-order lexical and semantic representations can feed back and mitigate the lower-level deficit. When the deficit is too severe, little top-down influence will be observed. Once the activation is sufficient, however, the more degraded the input representation and the longer the time required for processing, the more time is available for top-down support to accumulate. We have suggested that, in the context of a cascaded, interactive system, it is possible to observe lexical and semantic effects simultaneously with a peripheral lesion that affects lower-level processing. We have also proposed that the observed reading pattern, including the sequential LBL reading itself, arises from the residual function of the normal reading system that probably involves both the left and right hemispheres and that there is no reason to invoke the right-hemisphere reading system as the primary mediator of the lexical and semantic effects.

REFERENCES

Ajax, E.T. (1967). Dyslexia without agraphia. *Archives of Neurology, 17,* 645–652.

Albert, M.L., Yamadori, A., Gardner, H., & Howes, D. (1973). Comprehension in alexia. *Brain, 96,* 317–328.

Arguin, M., & Bub, D. (1993). Single character processing in a case of pure alexia. *Neuropsychologia, 31(5),* 435–458.

Arguin, M.E., & Bub, D. (1994a). Functional mechanisms in pure alexia: Evidence from letter processing. In M.J. Farah and G. Ratcliff (Eds.), *The neuropsychology of high-level vision* (pp. 149–171). Hillsdale, NJ: Lawrence Erlbaum Associates Inc.

Arguin, M.E., & Bub, D. (1994b). Pure alexia: Attempted rehabilitation and its implications for the interpretation of the deficit. *Brain and Language, 47,* 233–268.

Arguin, M.E., Bub, D., & Bowers, J.S. (this issue). Extent and limits of covert lexical activation in letter-by-letter reading. *Cognitive Neuropsychology, 15,* (1/2).

Baynes, K. (1990). Language and reading in the right hemisphere: Highways or biways or the brain. *Journal of Cognitive Neuroscience, 2,* 159–179.

Beeman, M., & Chiarello, C. (1997). *Right-hemisphere language comprehension: Perspectives from cognitive neuroscience.* Hillsdale, NJ: Lawrence Erlbaum Associates Inc.

Behrmann, M., Barton, J.J., & Black, S.E. (1998). *Eye movements reveal the sequential processing in letter-by-letter reading.* Manuscript in preparation.

Behrmann, M., Black, S.E., & Bub, D. (1990). The evolution of pure alexia: A longitudinal study of recovery. *Brain and Language, 39,* 405–427.

Behrmann, M., & McLeod, J (1995). Rehabilitation for pure alexia: Efficacy of therapy and implications for models of normal word recognition. *Neuropsychological Rehabilitation, 5,* 149–180.

Behrmann, M., Moscovitch, M., Black, S.E., & Mozer, M. (1990). Perceptual and conceptual factors in neglect dyslexia: Two contrasting case studies. *Brain 113(4),* 1163–1183.

Behrmann, M., Nelson, J., & Sekuler, E. (1998). *Visual complexity in letter-by-letter reading: "Pure" alexia is not so pure.* Neuropsychologia, in press.

Behrmann, M., & Shallice, T. (1995). Pure alexia: An orthographic not spatial disorder. *Cognitive Neuropsychology, 12(4),* 409–454.

Black, S.E., & Behrmann, M. (1994). Localization in alexia. In A. Kertesz (Ed.), *Localization and neuroimaging in neuropsychology* (pp. 331–376). San Diego, CA: Academic Press.

Bowers, J.S., Arguin, M., & Bub, D.N. (1996). Fast and specific access to orthographic knowledge in a case of letter-by-letter reading. *Cognitive Neuropsychology, 13(4)*, 525–567.

Bowers, J.S., Bub, D., & Arguin, M. (1966). A characterization of the word superiority effect in pure alexia. *Cognitive Neuropsychology, 13*, 415–441.

Brunn, J.L., & Farah, M. (1991). The relation between spatial attention and reading: Evidence from the neglect syndrome. *Cognitive Neuropsychology, 8(1)*, 59–75.

Bub, D., & Arguin, M. (1995). Visual word activation in pure alexia. *Brain and Language, 49*, 77–103.

Bub, D.N., Arguin, M., & Lecours, A.R. (1993). Jules Déjerine and his interpretation of pure alexia. *Brain and Language, 45*, 531–559.

Bub, D., Black, S.E., & Howell, J. (1989). Word recognition and orthographic context effects in a letter-by-letter reader. *Brain and Language, 36*, 357–376.

Bub, D., & Gum, T. (1988). *Psychlab experimental software.* Montreal: Montreal Neurological Institute.

Buxbaum, L., & Coslett, H.B. (1996). Deep dyslexic phenomenon in a letter-by-letter reader. *Brain and Language, 54*, 136–167.

Caplan, L.R., & Hedley-Whyte, T. (1974). Cuing and memory dysfunction in alexia without agraphia. *Brain, 97*, 251–262.

Chialant, D. & Caramazza, A. (this issue). Perceptual and lexical factors in a case of letter-by-letter reading. *Cognitive Neuropsychology, 15(1/2)*.

Cohen, J.D., MacWhinney, B., Flatt, M., & Provost, J. (1993). PsyScope: A new graphic environment for designing psychology experiments. *Behavioural Research Methods, Instruments and Computers, 25*, 257–271.

Coltheart, M. (1980). Deep dyslexia: A right hemisphere hypothesis. In M. Coltheart, K.E. Patterson, & J.C. Marshall (Eds.), *Deep dyslexia* (pp. 326–380). London: Routledge.

Coltheart, M. (1981). Disorders of reading and their implications for models of normal reading. *Visible Language, 15*, 245–286.

Coltheart, M. (1983). The right hemisphere and disorders of reading. In A.W. Young (Ed.), *Functions of the right cerebral hemisphere*, London: Academic Press.

Coslett, H.B., & Monsul, N. (1994). Reading with the right hemisphere: Evidence from transcranial magnetic stimulation. *Brain and Language, 46*, 198–211.

Coslett, H.B., & Saffran, E.M. (1989a). Evidence for preserved reading in "pure" alexia. *Brain, 112*, 327–359.

Coslett, H.B., & Saffran, E.M. (1989b). Preserved object recognition and reading comprehension in optic aphasia. *Brain, 112*, 1091–1110

Coslett, H.B. & Saffran, E.M. (1992). Optic aphasia and the right hemisphere: A replication and extension, *Brain and Language, 43*, 148–161.

Coslett, H.B., & Saffran, E.M. (1994). Mechanisms of implicit reading in alexia. In M.J. Farah & G. Ratcliff (Eds.), *The neuropsychology of high-level vision* (pp. 299–330). Hillsdale, NJ: Lawrence Erlbaum Associates Inc.

Coslett, H.B., Saffran, E.M., Greenbaum, S., & Schwartz, H. (1993). Reading in pure alexia. *Brain, 116*, 327–359.

Damasia, A., & Damasio, H. (1983). The anatomic basis of pure alexia. *Neurology, 33*, 1573–1583.

Doctor, E.A., Sartori, G., & Saling, M.M. (1990). A letter-by-letter reader who could not read nonwords. *Cortex, 26*, 247–262.

El Alaoui-Faris, M., Bienbelaid, F., Alaoui, C., Tahiri, L., Jiddane, M., Amarti, A., & Chkili, T. (1994). Alexie sans agraphie en langue arabie etude neurolinguistique et IRM. *Revue Neurologique (Paris), 150*, 771–775.

Farah, M.J. (1991). Patterns of co-occurrence among the associative agnosias: Implications for visual object recognition. *Cognitive Neuropsychology, 8(1)*, 1–19.

Farah, M.J. (1994). Visual perception and visual awareness after brain damage: A tutorial review. In C. Umiltà & M. Moscovitch (Eds.), *Attention and performance XV: Conscious and nonconscious information processing* (pp. 37–76). Cambridge, MA: Bradford Books, MIT Press.

Farah, M.J. (1997). Are there orthography-specific brain regions? Neuropsychological and computational investigations. In R.M. Klein & P.A. McMullen (Eds.), *Converging methods for understanding reading and dyslexia*. Cambridge, MA: MIT Press.

Farah, M.J., & Wallace, M. (1991). Pure alexia as a visual impairment: A reconsideration. *Cognitive Neuropsychology, 8(3/4)*, 313–334.

Feinberg, T.E., Dyckes-Berke, D., Miner, C.R., & Roane, D.M. (1995). Knowledge, implicit knowledge and metaknowledge in visual agnosia and pure alexia. *Brain, 118*, 789–800.

Freedman, L., & Costa, L. (1982). Pure alexia and right hemiachromatopsia in posterior dementia. *Journal of Neurology, Neurosurgery and Psychiatry, 55(6)*, 500–502.

Friedman, R.B., & Alexander, M.P. (1984). Pictures, images, and pure alexia: A case study. *Cognitive Neuropsychology, 1(1)*, 9–23.

Friedman, R.B., Beeman, M., Lott, S.N., Link, K., Grafman, J., & Robinson, S. (1993). Modality-specific phonological alexia. *Cognitive Neuropsychology, 10*, 549–568.

Friedman, R.B., & Hadley, J.A. (1982). Letter-by-letter surface alexia. *Cognitive Neuropsychology, 9*, 1–23.

Geschwind, N. (1965). Disconnection syndromes in animals and man. *Brain, 88*, 237–294.

Greenblatt, S.H. (1973). Alexia without agraphia. *Brain, 96*, 307–316.

Grossi, D., Fragassi, N.A., Orsini, A., Falco, F.A.D., & Sepe, O. (1984). Residual reading capability in a patient with alexia without agraphia. *Brain and Language, 23*, 337–348.

Hanley, J.R., & Kay, J. (1992). Does letter-by-letter reading involve the spelling system? *Neuropsychologia, 30*, 237–256.

Hanley, J.R., & Kay, J. (1996). Reading speed in pure alexia. *Neuropsychologia, 34(12)*, 1165–1174.

Henderson, L. (1982). *Orthography and word recognition in reading*. London: Academic Press.

Henderson, V., Friedman, R., Teng, E., & Weiner, J. (1985). Left hemisphere pathways in reading: Inferences from pure alexia without hemianopia. *Neurology, 35*, 962–968.

Hinton, G.E. (1990). Mapping part-whole hierarchies in connectionist networks. *Artificial Intelligence, 46*, 47–75.

Hinton, G.E., & Shallice, T. (1991). Lesioning an attractor network: Investigations of acquired dyslexia. *Psychological Review, 98(1)*, 74–95.

Horikoshi, T., Asari, Y., Watanabe, A., Nagaseki, Y., Nukui, H., Sasaki, H., & Komiya, K. (1997). Music alexia in a patient with mild pure alexia: Disturbed meaningful perception of nonverbal figures. *Cortex, 33*, 187–194.

Howard, D. (1991). Letter-by-letter readers: Evidence for parallel processing. In D. Besner and G.W. Humphreys (Eds.), *Basic processes in reading: Visual word recognition* (pp. 34–76). Hillsdale, NJ: Lawrence Erlbaum Associates Inc.

Johnston, J.C., & McClelland, J.L. (1980). Experimental tests of a hierarchical model of word identification. *Journal of Verbal Learning and Verbal Behavior, 19*, 503–524.

Just, M.A., & Carpenter, P. (1987). *The psychology of reading and language comprehension*. Boston, MA: Allyn & Bacon.

Karanth, P. (1985). Dyslexia in a Dravidian language. In K.E. Patterson, J.C. Marshall, & M. Coltheart (Eds.), *Surface dyslexia* (pp. 251–260). Hove, UK: Lawrence Erlbaum Associates Ltd.

Kashiwagi, T., & Kashiwagi, A. (1989). Recovery process of a Japanese alexic without agraphia. *Aphasiology, 3*, 75–91.

Kay, J., & Hanley, R. (1991). Simultaneous form perception and serial letter recognition in a case of letter-by-letter reading. *Cognitive Neuropsychology, 8(3/4)*, 249–273.

Kinsbourne, M. & Warrington, E.K. (1962). A disorder of simultaneous form perception. *Brain, 85*, 461–486.

Kreindler, A., & Ionescu, Y. (1961). A case of "pure" word blindness. *Journal of Neurology, Neurosurgery and Psychiatry, 24*, 275–280.

Kučera, F., & Francis, W.N. (1967). *Computational analysis of present-day American English*. Providence, RI: Brown University Press.

LaBerge, D. (1983). Spatial extent of attention and letters and words. *Journal of Experimental Psychol-*

ogy: Human Perception and Performance, 9, 371–379.

Làdavas, E., Umiltà, C., & Mapelli, D. (1997). Lexical and semantic processing in the absence of word reading: Evidence from neglect dyslexia. *Neuropsychologia, 35(8),* 1075–1085.

Landis, T., Regard, M., & Serrat, A. (1980). Iconic reading in a case of alexia without agraphic caused by a brain tumour. A tachistoscopic study. *Brain and Language, 11,* 45–53.

Lazar, R.M., & Scarisbrick, D. (1973). Alexia without agraphia: A functional assessment of behaviour in focal neurologic disease. *The Psychological Record, 43,* 639–650.

Levine, D.M., & Calvanio, R.A. (1978). A study of the visual defect in verbal alexia-simultanagnosia. *Brain, 101,* 65–81.

Marshall, J.C., & Newcombe, F. (1973). Patterns of paralexia: A psycholinguistic approach. *Journal of Psycholinguistic Research, 2,* 175–199.

Mayall, K., & Humpreys, G.W. (1996). A connectionist model of alexia: Covert recognition and case mixing effects: *British Journal of Psychology, 87,* 355–402.

McClelland, J.L. (1979). On the time relations of mental processes: An examination of systems of processes in cascade. *Psychological Review, 86,* 287–300.

McClelland, J.L. (1987). The case for interactionism in language processing. In M. Colthcart (Ed.), *Attention and Performance XII* (pp. 3–36). Hove, UK: Lawrence Erlbaum Associates Ltd.

McClelland, J.L., & Rumelhart, D.E. (1981). An interactive activation model of context effects in letter perception: Part 1. An account of basic findings. *Psychological Review, 88(5),* 375–407.

Michel, F., Henaff, M.A., & Intriligator, J. (1996). Two different readers in the same brain after a posterior callosal lesion. *NeuroReport, 7,* 768–788.

Miozzo, M., & Caramazza, A. (this issue). Varieties of pure alexia: The case of failure to access graphemic representations. *Cognitive Neuropsychology, 15(1/2).*

Monsell, S., Doyle, M.C., & Haggard, P.N. (1989). Effects of frequency on visual word recognition tasks: Where are they? *Journal of Experimental Psychology: General, 118(1),* 43–71.

Montant, M., Behrmann, M., & Nazir, T. (1998). *Lexical activation in pure alexia.* Manuscript in preparation.

Montant, M., Nazir, T.A., & Poncet, M. (this issue). Pure alexia and the viewing position effect in printed words. *Cognitive Neuropsychology, 15(1/2).*

Morton, J. (1969). The interaction of information in word recognition. *Psychological Review, 76,* 165–178.

Morton, J. (1979). Word recognition. In J. Morton & J.C. Marshall (Eds.), *Psycholinguistic Series 2.* London: Elek.

Mozer, M.C.. (1991). *The perception of multiple objects: A connectionist approach.* Cambridge, MA: MIT Press.

Mozer, M.C., & Behrmann, M. (1990). On the interaction of selective attention and lexical knowledge: A connectionist account of neglect dyslexia. *Journal of Cognitive Neuroscience, 2(2),* 96–123.

Mozer, M.C., & Behrmann, M. (1992). Reading with attentional impairments: A brain-damaged model of neglect and attentional dyslexia. In R. Reilly & N.E. Sharkey (Eds.), *Connectionist approaches to natural language processing* (pp. 409–460). Hove, UK: Lawrence Erlbaum.

Nelson, J., Behrmann, M., & Plaut, D.C. (1998). *The interaction of stimulus degradation and word length with frequency and with imageability.* Manuscript in preparation.

Nitzberg-Lott, S., Friedman, R.B., & Linebaugh, C.W. (1994). Rationale and efficacy of a tactile-kinaesthetic treatment for alexia. *Aphasiology, 8,* 181 195.

O'Regan, K., & Levy-Schoen, A. (1989). Eye-movement strategy and tactics in word recognition and reading. In K. Rayner & A. Pollatsek (Eds.), *The psychology of reading* (pp. 361–383). Englewood Cliffs, NJ: Prentice Hall.

Patterson, K.E., & Kay, J. (1982). Letter-by-letter reading: Psychological descriptions of a neurological syndrome. *Quarterly Journal of Experimental Psychology, 34A,* 411–441.

Perri, R., Bartolomeo, P., & Silveri, M.C. (1996). Letter dyslexia in a letter-by-letter reader. *Brain and Language, 53*, 390–407.

Plaut, D.C. (1995). Double dissociation without modularity: Evidence from connectionist neuropsychology. *Journal of Clinical and Experimental Neuropsychology, 17(2)*, 291–321.

Plaut, D.C. (1998). Connectionist modelling of normal and impaired word reading. Manuscript in preparation.

Plaut, D.C., McClelland, J.L., Seidenberg, M.S. (1995). Reading exception words and pseudowords: Are two routes really necessary? In J.P. Levy, D. Bairaktaris, J.A. Bullinaria, & P. Cairns (Eds.), *Connectionist models of memory and language* (pp. 145–159). London: UCL Press.

Plaut, D.C., McClelland, J.L., Seidenberg, M., & Patterson, K.E. (1996). Understanding normal and impaired word reading: Computational principles in quasi-regular domains. *Psychological Review, 103*, 56 115.

Plaut, D.C., & Shallice, T. (1993). Deep dyslexia: A case study of connectionist neuropsychology. *Cognitive Neuropsychology, 10(5)*, 377–500.

Price, C.J., & Humphreys, G.W. (1992). Letter-by-letter reading? Functional deficits and compensatory strategies. *Cognitive Neuropsychology, 9(5)*, 427–457.

Price, C.J., & Humphreys, G.W. (1995). Contrasting effects of letter-spacing in alexia: Further evidence that different strategies generate word length effects in reading. *Quarterly Journal of Experimental Psychology, 48A*, 573–597.

Rapcsak, S.Z., Rubens, A.B., & Laguna, J.F. (1990). From letters to words: Procedures for word recognition in letter-by-letter reading. *Brain and Language, 38*, 504–514.

Rapp, B.C., & Caramazza, A. (1991). Spatially determined deficits in letter and word processing. *Cognitive Neuropsychology, 8(3/4)*, 275–312

Rayner, K., & Pollatsek, A. (1987). Eye movements in reading: A tutorial review. In M. Coltheart (Ed.), *Attention and performance XII: The psychology of reading* (pp. 327–362). Hove, UK: Lawrence Erlbaum Associates Ltd.

Regard, M., Landis, T., & Hess, K. (1985). Preserved stenography reading in a patient with pure alexia. *Archives of Neurology, 42*, 400–402.

Reuter-Lorenz, P., & Brunn, J. (1990). A prelexical basis for letter-by-letter reading: A case study. *Cognitive Neuropsychology, 7*, 1–20.

Rumelheart, D.E., & McClelland, J.L. (1982). An interactive activation model of context effects in letter perception: Part 2. The contextual enhancement effect and some tests and extensions of the model. *Psychological Review, 89*, 60–94.

Saffran, E.M., Bogyo, L.C., Schwartz, M.F., & Marin, O.S.M. (1980). Does deep dyslexia reflect right hemisphere reading? In M. Coltheart, K.E. Patterson, & J.C. Marshall (Eds.), *Deep dyslexia* (pp. 381–406). London: Routledge.

Saffran, E.M., & Coslett, H.B. (this issue). Implicit vs. letter-by-letter reading in pure alexia: A tale of two systems. *Cognitive Neuropsychology, 15(1/2)*.

Schacter, D., McAndrews, M.P., & Moscovitch, M. (1988). Access to consciousness: Dissociations between explicit knowledge in neuropsychological syndromes: In L. Weiskrantz (Ed.), *Thought without language*. Oxford: Oxford University Press.

Schacter, D., Rapcsak, S.Z., Rubens, A.B., Tharan, M., & Laguna, J. (1990). Priming effects in a letter-by-letter reader depend upon access to the word form system. *Neuropsychologia, 28(10)*, 1079–1094.

Schiepers, C. (1980). Response latency and accuracy in visual word recognition. *Perception and Psychophysics, 27*, 71–81.

Seidenberg, M., & McClelland, J.L. (1989). A distributed, developmental model of word recognition and naming. *Psychological Review, 96*, 523–568.

Sekuler, E., & Behrmann, M. (1996). Perceptual cues in pure alexia. *Cognitive Neuropsychology, 13(7)*, 941–974.

Shallice, T. (1988). *From neuropsychology to mental structure*. Cambridge: Cambridge University Press.

Shallice, T., & Saffran, E. (1986). Lexical processing in the absence of explicit word identification:

Evidence from a letter-by-letter reader. *Cognitive Neuropsychology, 3(4),* 429–458.

Shallice, T., & Warrington, E.K. (1980). Single and multiple component central dyslexic syndromes. In M. Coltheart, K.E. Patterson, & J.C. Marshall (Eds.), *Deep dyslexia.* London: Routledge.

Sieroff, E., Pollatsek, A., & Posner, M.I. (1988). Recognition of visual letter strings following injury to the posterior visual spatial attention system. *Cognitive Neuropsychology, 5(4),* 427–449.

Snodgrass, J.G., & Mintzer, M. (1993). Neighbourhood effects in visual word recognition: Facilitatory or inhibitory? *Memory and Cognition, 21(2),* 247–266.

Snodgrass, J.G., & Vanderwart, M.A. (1980). A standardised set of 260 pictures: Norms for name agreement, image agreement, familiarity and visual complexity. *Journal of Experimental Psychology: Learning, Memory and Cognition, 6,* 174–215.

Speedie, L., Rothi, L.J., & Heilman, K.M. (1982). Spelling dyslexia: A form of cross-cuing. *Brain and Language, 15,* 340–352.

Sperling, G., & Melcher, M.J. (1978). Visual search, visual attention and the attention operating characteristic. In J. Requin (Ed.), *Attention and performance,* Vol. 7. Hillsdale, NJ: Erlbaum.

Stachowiak, F.L., & Poeck, K. (1976). Functional disconnection in pure alexia and color naming deficit demonstrated by facilitation methods. *Brain and Language, 3,* 135–143.

Staller, L., Buchanan, D., Singer, M., Lappin, J., & Webb, W. (1978). Alexia without agraphia: An experimental case study. *Brain and Language, 5,* 378–387.

Sternberg, S. (1969). The discovery of processing stages: Extensions of the Donders method. *Acta Psychologia, 30,* 276–315.

Strain, E., Patterson, K., & Seidenberg, M.S. (1995). Semantic effects in single-word naming. *Journal of Experimental Psychology: Learning, Memory & Cognition, 21(5),* 1140–1154.

Terry, P., Samuels, S.J., & LaBerge, D. (1976). The effects of letter degradation and letter spacing on word recognition. *Journal of Verbal Learning and Verbal Behavior, 15,* 577–585.

Townsend, J.T. (1990). Serial vs. parallel processing. *Psychological Science, 1,* 46–54.

Tuomainen, J., & Laine, M. (1991). Multiple oral re-reading technique in rehabilitation of pure alexia. *Aphasiology, 5,* 401–409.

Van Orden, G.C., Pennington, B.F., & Stone, G.O. (1990). Word identification in reading and the promise of subsymbolic psycholinguistics. *Psycholinguistic Review, 97(4),* 488–522.

Vargha-Kadem, F., Carr, L.J., Isaacs, E., Brett, F., Adams, C., & Mishkin, M. (1997). Onset of speech after left hemispherectomy in a 9-year-old boy. *Brain, 120,* 159–182.

Vigliocco, G., Semenza, C., & Neglia, M. (1992). Different deficits and different strategies in letter-by-letter reading. Paper presented at the Academy of Aphasia, Toronto.

Warrington, E.K., & Shallice, T. (1980). Word-form dyslexia. *Brain, 103,* 99–112.

Weekes, B.S. (1997). Differential effects of number of letters on word and nonword naming latency. *Quarterly Journal of Experimental Psychology, 50A(2),* 439–456.

Young, A.W., & Ellis, A.W. (1985). Different methods of lexical access for words presented in the left and right visual hemifields. *Brain and Language, 24,* 326–358.

Young, A.W., & De Haan, E. (1990). Impairments of visual awareness. *Mind and Language, 5,* 29–48.

EXTENT AND LIMITS OF COVERT LEXICAL ACTIVATION IN LETTER-BY-LETTER READING

Martin Arguin
Université de Montréal and Institut Universitaire de Gériatrie de Montréal, Montréal, Canada

Daniel Bub
Institut Universitaire de Gériatrie de Montréal, Montréal and University of Victoria, Canada

Jeffrey Bowers
Rice University, Houston, USA

The occurrence of implicit reading in brain-damaged patients with letter-by-letter dyslexia suggests a process of covert lexical activation, whereby lexical access occurs on the basis of parallel letter encoding. The extent and limitations of this process were studied by examining masked orthographic and phonological word priming as well as orthographic neighbourhood size effects in letter-by-letter reader IH. In Exp. 1, masked repetition priming occurred with primes displayed in a case-alternate format that were shown for 100 msec (a duration that does not reliably support overt word identification in IH). Under similar exposure conditions, however, primes that are homophones to the target failed to affect performance, in contrast to neurologically intact observers (Exp. 2). Exp. 3 showed that IH's naming latencies are reduced for words with many (vs. few) orthographic neighbours. This result suggests that overt word recognition in the patient is not strictly mediated by sequential letter recognition, but rather that it is conjointly affected by covert lexical activation. Relative to neurologically intact subjects, however, the pattern of the neighbourhood size effect shown by IH as a function of word frequency is abnormal and suggests that lexical activation based on the parallel processing of letters is weakened in the patient compared to normal readers. Overall, results from IH point to a weak form of activation of abstract orthographic lexical representations on the basis of parallel letter encoding, but no significant degree of phonological access. This account is discussed in relation to other similar proposals seeking an explanation of letter-by-letter dyslexia and of the covert lexical activation phenomena that accompany the disorder.

Requests for reprints should be addressed to Martin Arguin, PhD, Département de psychologie, Université de Montréal, CP 6128, Succ. Centre-ville, Montréal, Qc, Canada, H3C 3J7. Tel: 514-343-2167; Fax: 514-343-5787; e-mail: arguinm@ere.umontreal.ca.

We are grateful to IH for his collaboration in this project. This research has been supported by a grant from the Medical Research Council of Canada to the first author. Martin Arguin is chercheur-boursier of the Fonds de la Recherche en Santé du Québec.

INTRODUCTION

Letter-by-letter (LBL) dyslexia is an acquired reading disorder that typically follows from a left occipital lesion (Damasio & Damasio, 1983; Déjerine, 1891) and is therefore usually accompanied by right hemianopia. As its name implies, the disorder is characterised by behavioural manifestations suggesting that reading is effected in a serial, letter-by-letter manner, in contrast to the parallel process observed in neurologically intact readers (e.g. Henderson, 1982, for review). Thus, the time required to read a word aloud increases from 500 msec to several seconds (depending on the patient) for each additional letter in the stimulus (e.g. Arguin & Bub, 1993; Bowers, Bub & Arguin, 1996; Farah & Wallace, 1991; Patterson & Kay, 1982; Reuter-Lorenz & Brunn, 1990; Warrington & Shallice, 1980). Consequently, single-word reading latencies are far above those found in normal observers and LBL patients report that reading has become a tedious and very effortful act.

The kind of functional damage that may be held responsible for the clinical symptoms of LBL reading is still a controversial topic. To summarise briefly, here are the main accounts proposed so far for LBL reading: (1) poor perceptual encoding (Farah & Wallace, 1991; Friedman & Alexander, 1984; Kinsbourne & Warrington, 1962; Levine & Calvanio, 1978; Rapp & Caramazza, 1991); (2) deficit in abstract letter identification (Arguin & Bub, 1993, 1994; Behrmann & Shallice, 1995; Kay & Hanley, 1991; Reuter-Lorenz & Brunn, 1990); (3) impaired transfer of letter identities to global orthographic word-forms (Patterson & Kay, 1982); (4) damaged orthographic word-form system (Warrington & Shallice, 1980); (5) impaired access to phonological word-forms following a relatively intact access to orthographic word-forms (Bowers, Arguin, & Bub, 1996). Yet other authors have suggested that LBL reading may not be a unitary syndrome and that, ultimately, the functional impairments causing the disorder may be as varied as the LBL readers themselves (Price & Humphreys, 1992, 1995).

Of interest, and in contrast to the slow sequential letter identification procedure that seems necessary for overt word recognition in LBL patients, a minority of cases paradoxically show evidence suggesting accurate and rapid parallel lexical access for words they cannot identify explicitly. In the first published report of this phenomenon called *implicit reading*, Shallice and Saffran (1986) showed that lexical decisions or semantic decisions in an LBL patient can be performed with an accuracy level that is above chance, even though exposure durations were too short to allow for explicit identification of the stimuli. Similar demonstrations were also reported by Coslett and Saffran (1989) and by Coslett, Saffran, Greenbaum, and Schwartz (1993). Coslett and Saffran (1989) have also shown in two LBL readers that, under limited exposure duration conditions, error rates in the lexical decision task may be independent of word length, in contrast to overt recognition performance. In a more recent study, Bub & Arguin (1995) reported that accurate lexical decisions by an LBL reader can be carried out with response latency being unaffected by word length. It thus appears that, although overt word recognition performance may require the serial identification of letters, lexical or semantic classification in some LBL cases may occur on the basis of parallel letter encoding.

Another demonstration of implicit reading in LBL dyslexia has come from the word superiority effect. In normal observers, the recognition of a briefly exposed letter that is backward masked is superior if it is part of a word than if it is shown in isolation or if it is part of a random letter string (Johnston, 1978; McClelland & Johnston, 1977; Reicher, 1969; Wheeler, 1970). This finding has been reported in some LBL readers, again with exposure durations too short to allow for overt stimulus identification (Bub, Black, & Howell, 1989; Reuter-Lorenz & Brunn, 1990). As with the lexical and semantic classification results, the word superiority effect in LBL readers has been considered as suggesting a rapid access to orthographic word-forms that is mediated by a process other than the serial letter identification involved in explicit word recognition.

What the occurrence of implicit reading suggests is that, besides the slow letter-by-letter process LBL patients seem to require to recognise a word consciously, they may also have access to a lexical access procedure that operates much more rapidly and in parallel, but which cannot reliably support explicit word recognition on its own. We will refer to this putative reading procedure in LBL reading under the term of covert lexical activation.

Covert lexical activation may be assumed to reflect the residual operation of the reading system that has been damaged by the brain lesion (Behrmann, Plaut, & Nelson, this issue; Bub & Arguin, 1995; Bub et al., 1989; Montant, Nazir, & Poncet, this issue; Shallice & Saffran, 1986) or the implication of a separate system that does not provide a significant contribution to reading in normals (Buxbaum & Coslett, 1996; Coslett & Saffran, 1989; Coslett et al., 1993; Saffran & Coslett, this issue). In either case, a detailed characterisation of the extent and limits of covert lexical activation and of the factors that affect its occurrence or magnitude appear crucial for an accurate understanding of LBL reading. Using evidence from implicit reading to obtain such a specification has proven difficult, however, Possibly the most significant obstacle in using implicit reading as a probe into the impaired reading mechanism of LBL readers is the fact that the phenomenon fails to occur in many patients (Behrmann, Black, & Bub, 1990; Behrmann & Shallice, 1995; Howard, 1991; Patterson & Kay, 1982; Price & Humphreys, 1992, 1995; Warrington & Shallice, 1980). From this, one might simply conclude that many LBL patients have no covert lexical activation. Alternatively, this failure may depend on the unusual demands of the implicit reading task rather than on the absence of covert lexical activation. Indeed, by definition, the implicit reading task requires subjects to produce an overt decision about a stimulus they are unable to recognise consciously. As in blindsight, patients may reasonably be reluctant in producing this sort of response, and this would prevent any manifestation of implicit reading altogether. Thus, reasons other than a failure of covert lexical activation may possibly explain past difficulties in demonstrating implicit reading in LBL patients. A similar reasoning has been proposed by Coslett et al. (1993), who showed that the particular strategy adopted by the patient may determine the occurrence of implicit reading.

Besides the implicit reading approach, another way the issue of covert lexical activation may possibly be addressed is through the performance of LBL patients in overt word recognition tasks. Thus, even if serial letter identification appears obligatory for overt word recognition, it remains possible that covert lexical activation may contribute to this performance. Thus, in addition to receiving inputs from a serial letter identification mechanism, the representation system mediating overt word recognition in LBL reading may also receive inputs from that involved in covert lexical activation. Alternatively, overt word recognition in LBL readers may be mediated by the same system as that involved in covert lexical activation. In this case, letter identity information obtained through a rapid parallel process would allow covert lexical activation effects, but this activation would need to be supplemented by serial letter identification in order to permit over word recognition.

The first evidence suggesting that overt word recognition in LBL dyslexia may not always be strictly based on serial letter identification was provided by Howard (1991; but see Behrmann & Shallice, 1995, for discrepant findings). He showed that "fast" reading responses in a visual word naming task resulted from the parallel processing of the constituent letters in the target, but that this process was subject to a significant rate of error. Only when this process failed did patients have to resort to serial letter identification for the overt recognition of a word. This suggests that the lexical activation process assumed to be responsible for implicit reading when it remains covert, may in fact become overt on some proportion of trials, thereby allowing the patient to recognise a word without serial letter processing. When this serial processing is required, however, Howard's results do not indicate whether lexical activation resulting from parallel letter encoding has any contribution to overt word recognition performance.

More recent studies by our group have shown that covert lexical activation may affect overt word recognition performance in an LBL reader. The patient examined in those studies is IH, who suffers from LBL surface dyslexia (Friedman & Hadley, 1992). In a task where the patient was asked to read overtly (i.e. full report) letter strings that were displayed briefly (83 msec) and then masked, recognition accuracy was higher when the stimulus was a word than when it was a nonword (Bowers, Bub, et al., 1996). This lexical effect on overt recognition performance occurred with words and nonwords matched pairwise on orthographic regularity, and a separate experiment discounted an explanation of the results based on guessing. It appears unlikely that these observations can be explained by assuming that explicit stimulus identification was based exclusively on the serial processing of individual letters. Rather, it was proposed that some form of parallel or global encoding of the letter string—i.e. covert lexical activation—must have occurred in IH to allow better identification of the letter string if it formed a word than a nonword.

A separate study examined the effect of briefly exposed primes on IH's overt word recognition performance (Bowers, Arguin, et al., 1996). The subject was shown target words, printed in upper-case, which had to be read aloud. Targets were preceded by a 100 msec lower-case prime and by a 17 msec pattern mask, each displayed in sequence. Items were made of letters that greatly changed shapes between upper- and lower-case formats so that any priming effect could not be a function of the physical overlap between prime and target. Rather, priming effects under such conditions would imply a fast abstract orthographic encoding of the prime[1]. In one experiment, primes were either the same word as the subsequent target or an unrelated word. Correct response times (RTs) were much shorter for targets preceded by an identity prime than by an unrelated prime. A separate experiment showed that this priming effect is highly specific. Indeed, primes that were orthographic neighbours to the target (i.e. words of the same length as the target and differing from it by just one letter; Coltheart, Davelaar, Jonasson, & Benner, 1977) failed to result in any performance benefit relative to unrelated primes. This was true whatever the letter position by which neighbour-primes differed from the target. These priming effects on overt word recognition were obtained under prime exposure conditions (backward masked 100msec exposure) that do not reliably support overt recognition in the patient (see Bowers, Bub, et al., 1996). The results therefore again point to covert lexical activation affecting overt word recognition performance in IH.

The purpose of the present paper is to further study covert lexical activation in LBL reading and to try to characterise it in some detail. As in Bowers, Bub, et al. (1996) and Bowers, Arguin, et al. (1996), we used overt word recognition performance as the measure for covert lexical activation effects. In two experiments, the word priming paradigm of Bowers, Arguin, et al. (1996) served in attempts to establish boundary conditions for covert lexical activation in LBL reading. In a third experiment, the effect of orthographic similarity of the target to other words of the vocabulary was used as an index of covert lexical activation.

CASE REPORT

The patient who took part in the present experiments is IH. The word superiority and abstract word priming experiments of Bowers (Bowers, Arguin, et al., 1996; Bowers, Bub, et

---

[1] The occurrence of priming under these conditions may also conceivably reflect activation of the phonological, semantic, or other high-level representations of the prime. However, it is assumed here that such access from a visual written input must be mediated by an internal orthographic representation of the stimulus. To avoid overestimating the reading capacities suggested by priming effects in an LBL patient, we will therefore only assume access to the lowest level of representation that may mediate the relation between the prime and the target as the cause of priming.

al., 1996) that are described in the previous paragraphs have been conducted on IH, and details of his clinical status can be found in those publications. We will therefore only briefly summarise his condition. IH is a right-handed English-speaking male who was 56 years of age at the time of testing. At the age of 43, in 1983, IH suffered from a subarachnoid haemorrhage that was drained surgically. No CT or MRI scan is available but the neurological report indicates that the haematoma was located in the left temporo-occipital area. IH's behavioural complaints are of a complete right-homonymous hemianopia, anomia, surface agraphia, and reading problems. The patient's reading latencies average at about 1200–1500msec for 4-letter words and increase linearly by about 500msec each additional letter in the word. Therefore, the patient shows the characteristic clinical symptoms of LBL reading. IH's reading performance is also affected by lexical frequency and by the regularity of spelling-to-sound correspondences. Thus, word naming accuracy for regular words average about 85% correct across a vast range of word frequencies and this latter variable had no effect on accuracy. In contrast, accuracy on irregular words was of 69% correct with high-frequency items but performance dropped to 31% correct for low-frequency items. Such an interactive effect of frequency and regularity has been reported previously in surface dyslexics and suggests that IH suffers from a combination of LBL reading and surface dyslexia, a disorder called letter-by-letter surface dyslexia by Friedman and Hadley (1992).

EXPERIMENT 1

Bowers, Arguin, et al. (1996) have shown that a four-letter upper-case target word, preceded by the same word printed in lower-case, displayed for 100msec, and then masked, results in marked RT reductions compared to targets preceded by an unrelated prime. In those experiments, there was little overlap between the visual features of the lower-case prime and the upper-case target, which suggested that the priming effect was mediated by an orthographically abstract covert lexical activation procedure. Just how abstract this covert lexical activation is remains to be determined, however.

One hypothesis is that the covert lexical activation that mediates priming corresponds to the orthographic encoding process generally assumed to mediate normal reading. By this hypothesis, each letter of the prime is encoded in parallel as an abstract orthographic identity and these letter identities are transferred to an abstract lexical representation system. Thus, it would be the activation of the abstract orthographic representation of the prime that is responsible for the beneficial effect of identity priming on recognition of the target word. A viable alternative hypothesis, however, is that priming is mediated by shape-specific lexical knowledge. According to this view, the patient would have access to stored representations of word shapes under their lower-case and upper-case formats and the activation of a word in one of these formats by the prime would then transfer to the representation of the same word under the other format (see Boles, 1992;

Boles & Eveland, 1983, for a similar proposal for abstract letter recognition). This activation transfer between shape-specific representations of the same word would therefore be the process responsible for the priming effects previously observed in IH.

One way to distinguish between abstract orthographic vs. shape-specific word encoding as the process mediating word priming in LBL dyslexia is to present primes printed in an alternation of upper- and lower-case letters. According to the hypothesis that covert lexical activation is based on a shape-specific reading mechanism, the occurrence of priming depends on the prior existence of stored shape-specific word representations that match the surface features of the prime and the target that are presented. Words printed in a case-alternated format are not part of the normal reading environment and thus should not have prior shape-specific representations. Using such words as primes therefore should not result in any priming effect if covert lexical activation is based on a shape-specific lexical code. In contrast, the priming effect should still occur with case alternated primes if covert lexical activation is mediated by a truly abstract orthographic encoding operation. It has been reported by Forster and Guess (1996) that masked priming effects are unaffected by the case alternation manipulation in neurologically intact observers.

The contrast between the rival hypotheses of abstract orthographic vs. shape-specific word encoding to account for the word priming results in IH relates to recent findings about the left- and right-hemisphere orthographic encoding mechanisms that mediate reading. In a word repetition priming task conducted with neurologically intact observers, Marsolek and his collaborators (Marsolek, Kosslyn, & Squire, 1992; Marsolek, Squire, Kosslyn & Lulenski, 1994) have reported greater priming with displays to the right than to the left hemisphere if stimulus shape remained constant between study and test. However, if stimulus shape was changed (upper-case vs. lower-case print) between study and test, the priming effect was reduced with right-hemisphere stimulation to become equal to that for the left hemisphere, which was unaffected by the shape change manipulation. From these observations, the authors concluded that two separate systems contribute to visual word recognition, one that is abstract with respect to visual shape and another that is shape specific. Further, although both hemispheres would implement an abstract orthographic system, it was proposed that the shape-specific system operates more effectively in the right than the left hemisphere. This interpretation, it should be noted, implies that the full magnitude of the priming effects observed is exclusively attributable to the hemisphere to which stimuli were directed in the test phase. This need not be so, however, since neurologically intact subjects are capable and likely to transfer information between their cerebral hemispheres in a reading task where stimuli are lateralised. Assuming such transfer may have occurred, the alternative hypothesis of exclusive capacities for abstract orthographic encoding and for shape-specific encoding in the left and right cerebral hemispheres, respectively, is just as consistent with the data as the interpretation suggested by Marsolek and collaborators. By this alternative account, information about right-hemisphere stimuli was transferred early to the left hemisphere after an initial shape-specific encoding. While this right-hemisphere shape-specific mechanism would be responsible for the greater priming effect with right-hemisphere stimuli when shape remained constant between study and test, all the other components of the priming effects observed would be attributable to the abstract mechanism of the left hemisphere. A separate set of observations from a subject in whom inter-hemispheric transfer was impossible argues for our alternative interpretation of the Marsolek et al. data.

Reuter-Lorenz and Baynes (1992) studied split-brain patient JW in a task comparing the effects of abstract and physical identity primes on the recognition of a subsequent lateralised target letter. In this experiment, JW showed benefits from abstract and physical identity priming with left-hemisphere stimuli, but only physical identity priming with right-hemisphere stimuli. These observations suggest that, without the benefit of inter-hemispheric transfer, written stimuli exposed to the right hemisphere are represented under a shape-specific code only[2]. It seems clear that further work will be requited to elucidate fully the issue of hemispheric asymmetries in orthographic representation. However, to the degree that the available relevant data can be considered meaningful, it appears the conditions of the present experiment may help determine the lateralisation of covert lexical activation in IH. Thus, the lack of a priming effect with case-alternated primes in IH would strongly suggest that covert lexical activation is mediated by the right hemisphere. In contrast, a substantial priming effect under the conditions of Exp. 1 would suggest that the left hemisphere may be largely responsible for covert lexical activation effects in the patient.

This issue of hemispheric asymmetries in reading mechanisms is particularly relevant in the study of covert lexical activation in LBL readers. Indeed, resolution of this issue would indicate whether covert lexical activation is

---

[2] One may conceive this suggestion as being contradictory to observations by Saffran (Saffran, 1980; Saffran & Marin, 1977); they reported accurate reading performance with case alternated words in deep dyslexic patients, who are assumed to recognise words via their right hemispheres. The contradiction may be more apparent than real, however. Indeed, none of the deep dyslexic patients examined in those studies had a complete left hemianopia. In fact, three out of the four cases reported had normal visual fields whereas the other showed a right upper quadrant defect. Taken in conjunction with the fact that stimuli were presented in free vision, it is quite conceivable that abstract orthographic encoding of the words may have been performed by the left hemisphere and that the contribution of the right hemisphere in the reading performance of these patients only emerged at the stage of lexical access. In defence of this view, it may also be noted that all the signs suggesting a contribution of the right hemisphere in the reading performance of deep dyslexics concern the form of the lexical representations instantiated by that hemisphere, not the mechanisms involved in orthographic encoding.

based on the residual function of the system that mediated reading prior to the occurrence of brain damage (i.e. left hemisphere), or whether it implicates another system that contributes little to reading performance in neurologically intact individuals (i.e. right hemisphere). Studies by Coslett, Saffran, and their collaborators have suggested that the residual reading abilities of LBL readers may be mediated by the right hemisphere. Thus, high-imagery words tend to be read better than low-imagery words and concrete nouns tend to be read better than function words (Coslett & Saffran, 1989; Coslett et al., 1993). Furthermore, in a more recent experiment, transcranial electromagnetic stimulation applied over the posterior portion of the right hemisphere was found to disrupt overt word recognition in an LBL patient (Coslett & Monsul, 1994). This sort of evidence has led to the proposal that the putative right hemisphere contribution to word recognition in LBL patients may extend to implicit reading as well. Resolution of the issue of hemispheric contributions to covert lexical activation is crucial for our understanding of the phenomenon and for the design of rehabilitation attempts for the disorder.

In Exp. 1, we contrasted the rival hypotheses described earlier as to the process mediating covert lexical activation. The word priming procedure used was similar to that of Bowers, Arguin, et al. (1996). However, rather than preceding the upper-case target by a lower-case prime, primes in the present experiment were printed in an alternation of lower- and upper-case letters.

Methods

Subjects were IH and a group of 8 neurologically intact individuals (3 males, 5 females) aged between 18 and 20 years. The latter subjects served to determine that the results expected from an intact mature reading system actually occurred under the experimental conditions that were used. All trials began with a 1500msec rectangular 1.0cm (1.1° of visual angle, from a viewing distance of about 50cm) high × 3.5cm (4.0°) wide pattern mask made of a chequerboard with 1mm black and white elements, which was displayed at the centre of the computer screen. Subjects were requested to keep their eyes fixated at the rightmost extremity of the chequerboard. This procedure was required to ensure that the entire length of the primes and targets was within the normal portion of IH's visual field. All stimuli that followed the initial mask were also centred on the middle of the screen and the primes and targets all had vertical and horizontal extents that were inferior to those of the mask. Immediately following (i.e. 0msec interstimulus interval) the initial mask, a 100msec prime-word was displayed. It was then immediately followed by the pattern mask shown for 17msec, which immediately preceded target exposure. Targets ($N = 60$; Appendix A) were 4-letter upper-case words displayed within a rectangular 1.0cm (1.1°) high × 3,5cm (4.0°) wide frame and they remained visible until response. The subject was instructed to name the target aloud as rapidly as possible while avoiding errors. Half of the targets were high-frequency words (range 50–761 occurrences per million; average

= 239; Kucera & Francis, 1967) and the other half were low-frequency words (range: 3–20 occurrences per million; average = 9). High- and low-frequency targets were matched pair-wise on single-letter (average sum across words = 2100.5) and bigram frequencies (average sum across words = 211.8; Mayzner & Tresselt, 1965) and on their numbers of orthographic neighbours (average = 9.9; words of the same length as the target that differ from it by a single letter; Coltheart et al., 1977). Targets were selected so that at least three of their component letters had very different shapes between upper-case and lower-case formats (a/A, b/B, d/D, e/E, g/G, l/L, m/M, q/Q, r/R; Boles & Clifford, 1989). We also attempted to avoid targets with irregular spelling-to-sound correspondences because of the patient's difficulty in reading such stimuli aloud. Overall, the target list comprised three high-frequency and two low-frequency targets that were irregular. Primes were printed in an alternation of upper- and lower-case letters and were either the same word as the target (Repeated) or a different word with no orthographic overlap with the target (Unrelated). Unrelated primes were taken from the same frequency range as the target and each target was tested under both priming conditions. The first letter of half the primes in each condition was lower-case and the other half began with an upper-case letter. All stimuli appeared in black over a white background. Written stimuli were printed in Helvetica 24-point bold font. Responses were registered by a voice-key connected to the computer controlling the experiment. After each response, the experimenter registered the subject's utterance via the computer keyboard and then triggered the next trial by a keypress. To ensure enough observations per condition, each subject was administered the complete set of prime-target pairs twice. Across administrations, the case alternation of primes was inverted such that, for instance, the prime word "band" was printed "BaNd" on one administration and "bAnD" on the other. In IH, these administrations were conducted in different sessions separated by an interval of 2 weeks. In normal subjects, the order in which the lists were repeated was counterbalanced. Throughout the experiment with IH, a total of 4 trials (1.6%) were lost due to the failure of the subject's response to trigger the microphone. Across all trials run with neurologically intact subjects, 24 trials (1.3%) were lost due to a microphone error. These trials were not considered in the data analyses.

Results

Average correct response times (RTs) and error rates observed in neurologically intact subjects are shown in Figs. 1 and 2, respectively. The correlation between RTs and error rates was of – 0.26 (n.s.), which indicates no speed-accuracy trade-off. For each subject, RTs that were more than three SDs away from the mean for their condition were discarded. A total of 49 data points (2.6% of trials) were removed from the analysis on this criterion. A two-way ANOVA of Priming (Repeated vs. Unrelated) × Frequency (Low vs. High) showed main effects of priming [$F(1, 7) = 25.0; P < .005$] and a

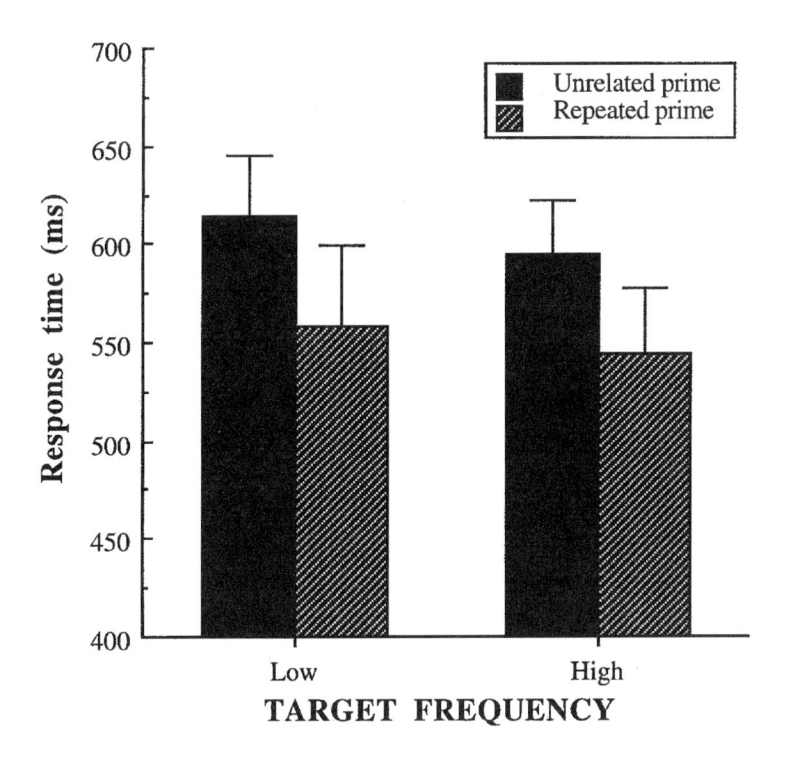

Fig. 1. Average correct RTs to low- and high-frequency targets preceded by unrelated or repeated case-alternated primes in neurologically intact subjects (Exp. 1).

marginally significant effect of frequency [$F(1, 7)=5.0; P < .07$], but no interaction [$F(1, 7) < 1$]. The main effect of priming indicates shorter RTs in the Repeated than in the Unrelated condition and the trend for a frequency effect suggests shorter RTs with high- than low-frequency targets. The ANOVA applied on error rates showed no significant effect of priming or of frequency, and no interaction [all $F's(1, 7) < 1$].

For IH, average correct RTs are shown in Fig. 3 and error rates are presented in Fig. 4. The correlation between RTs and error rates was of -0.17 (n.s.), which indicates no speed-accuracy trade-off. Four data points (1.7% of trials) were removed from the RT analysis because response latencies were more than three SDs away from the mean for their condition. A two-way ANOVA of Priming (Repeated vs. Unrelated) × Frequency (Low vs. High) showed a main effect of priming [$F(1, 176) = 10.1; P < .005$] but no main effect of frequency [$F(1, 176) = 1.5$; n.s.] and no interaction [$F(1, 176) = 2.7$; n.s.]. The main effect of priming indicates shorter RTs in the Repeated than in the Unrelated condition. Analysis of error rates showed no effect of priming [$\chi^2(1) = 1.6$; n.s.], but a higher error rate with low- than with high-frequency targets [$\chi^2(1) = 5.1; P < .05$].

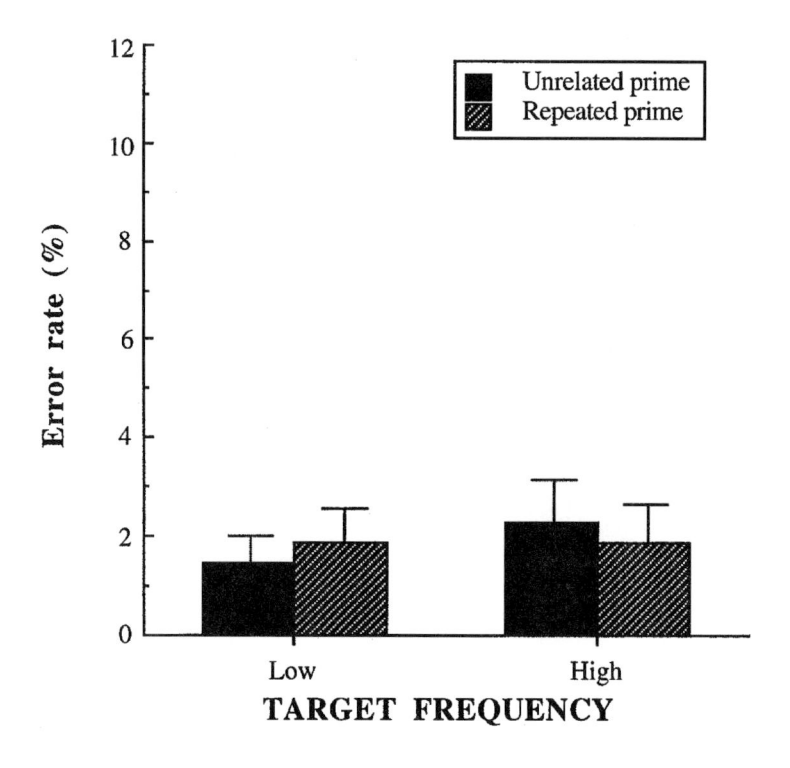

Fig. 2. Percentage error rates to low- and high-frequency targets preceded by unrelated or repeated case-alternated primes in neurologically intact subjects (Exp. 1).

Discussion

The results of Exp. 1 show that a word prime printed in a case-alternated format and displayed for 100msec markedly affects reading latency for an upper-case target word both in neurologically intact observers and in an LBL reader. Specifically, reading latency was reduced if the prime was the same word as the target rather than an unrelated word. The observations from IH replicate the abstract word priming effect, previously reported by Bowers, Arguin, et al. (1996) with primes printed in lower-case and targets in upper-case letters. Although the absolute magnitude of the priming effect observed in Exp. 1 in IH is substantially larger than in normal subjects, the size of priming effects relative to overall average correct RTs is highly similar, with an effect size of 10.0% for IH and of 10.9% for normals. The indication is thus that IH was as sensitive to the case-alternated primes as were neurologically intact subjects.

There are indications that the priming effect observed in IH was not mediated by the overt recognition of the prime. In IH, the display of a word for 100msec, which is then followed by a pattern mark, is insufficient to reliably support overt identification. Thus, with 133msec masked exposure, IH's word recognition accu-

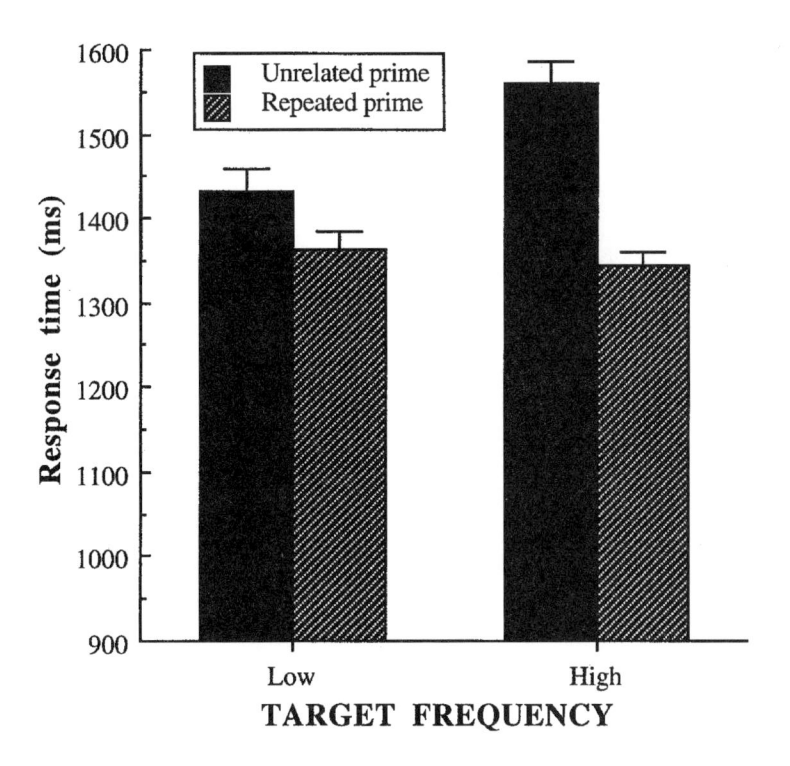

Fig. 3. Average correct RTs to low- and high-frequency targets preceded by unrelated or repeated case-alternated primes in IH (Exp. 1).

racy is only about 30% (Bowers, Bub, et al. 1996). In addition, IH never spontaneously reported seeing anything prior to the target and, when asked by the experimenter, he indicated he only occasionally saw a brief flash but that he had no idea what it could be. In spite of this, however, it could still be argued that there may have been a small proportion of trials on which IH was able to consciously recognise the prime and that only these trials are responsible for the priming effect observed. On this view, the failure of the patient to report even seeing the prime could be explained by a problem with memory, not perception. What the argument would predict, though, is that the distribution of IH's correct RTs with repeated primes should be bimodal. Thus, one portion of the RT distribution should comprise a number of very short response latencies corresponding to trials where the prime was consciously recognised and priming occurred. These trials should be segregated from the remainder, where RTs are much longer because the prime was not recognised consciously and therefore that no priming occurred. An analysis of the response latency distribution with repeated primes for IH fails to support this prediction of a bimodal distribution. Figure 5 shows the histogram of IH's actual RT distribution with repeated primes against the log-normal distri-

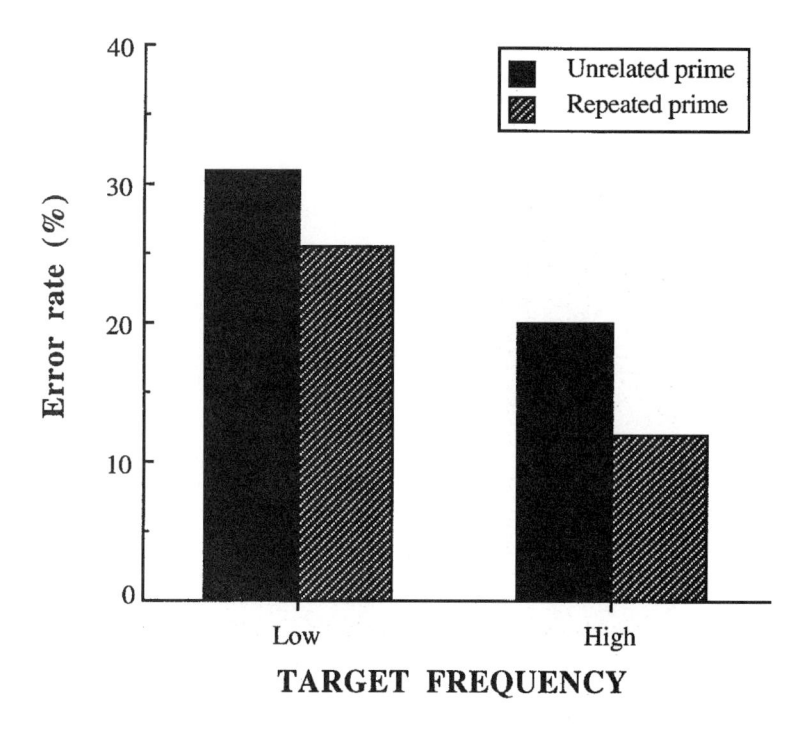

Fig. 4. Percentage error rates to low- and high-frequency targets preceded by unrelated or repeated case-alternated primes in IH (Exp. 1).

bution. Both distributions are highly similar and a Kolmogorov-Smirnov test indicates no difference between the two (D = 0.08; $P > .20$). Most importantly, the RT distribution shown by IH is quite distinct from the bimodal distribution that is predicted by the assumption that any priming effect observed in the patient is only due to the overt recognition of some proportion of the primes.

The central motivation for Exp. 1 was to determine the kind of orthographic encoding procedure on which covert lexical activation is based in IH. Results from Bowers, Arguin, et al. (1996) indicated that the priming effect mediated by covert lexical activation is orthographically abstract since it occurred with primes printed in lower-case and target printed in upper-case and with items made of letters that greatly change shapes across case. Still, as indicated earlier, priming under those conditions could have been based on shape-specific lexical codes rather than on the abstract encoding of letter identities. With primes printed in a case-alternated format as in the present experiment, the former theory predicted no priming effect since it may be reasonably assumed that shape-specific lexical representations for case-alternated words are not available to the patient. Contrary to this prediction, robust priming was found with case-alternated primes. This suggests that covert lexical activation in IH is based on an

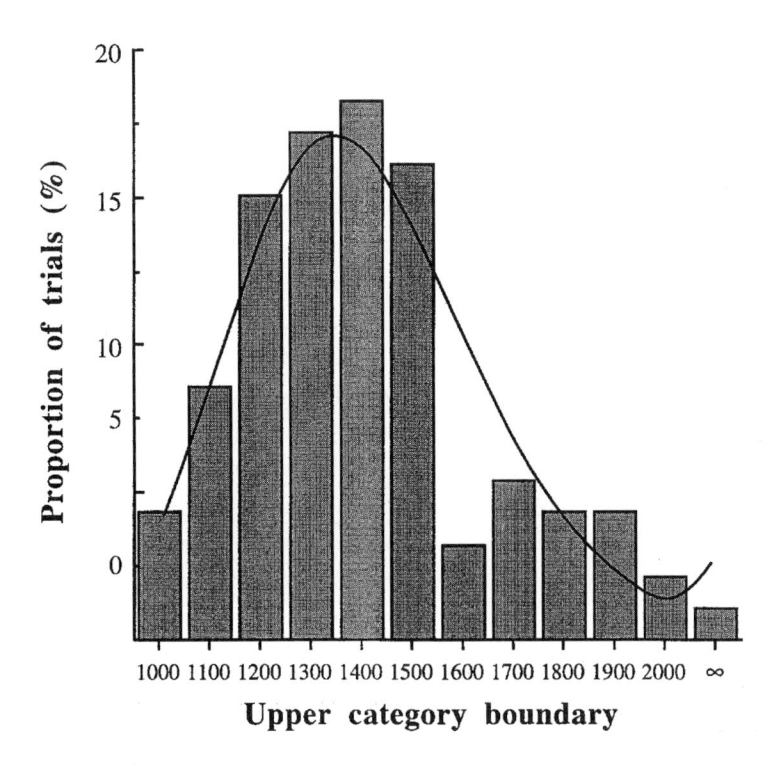

Fig. 5. Distribution of IH's correct RTs with repeated primes (histograms) against the log-normal distribution (continuous curve).

abstract orthographic encoding procedure, whereby the identity of each letter of the stimulus is determined rapidly while discarding visual shape information. This form of orthographic encoding corresponds to what is generally assumed to occur in normal readers, thus suggesting that covert lexical activation in IH may rest on the residual function of the system on which reading was based prior to the occurrence of brain damage.

Another issue of concern for Exp. 1 was that of the cerebral lateralisation of the covert lexical activation procedure assumed to mediate the word priming effect as well as implicit reading phenomena in LBL dyslexia. According to the interpretation presented earlier for the results of studies by Reuter-Lorenz and Baynes and by Marsolek and his collaborators, the left cerebral hemisphere is clearly dominant in performing abstract orthographic encoding of written stimuli whereas the right hemisphere would mainly, if not exclusively, rely on shape-specific representations. The kind of abstract encoding denoted by the priming effects shown by IH seems incompatible with covert lexical activation being mediated by the assumed shape-specific representation system of the right hemisphere. Rather, given what is presently known of the orthographic encoding capacities of the left and right cerebral hemispheres of split-brains and of neurologically intact individuals, the present results appear more compatible with the hypothesis that covert lexical activation in IH depends on the residual component of the left hemisphere's abstract orthographic encoding mechanism. The diagnostic criterion used here to identify the hemisphere responsible for covert lexical activation effects in LBL reading is new and of potential interest for further investigations. It should be acknowledged, however, that the hypothesis of covert lexical activation being mediated by the left cerebral hemisphere presently rests on relatively weak ground and that a firm conclusion regarding this issue must await further investigations of the relative capacities of the cerebral hemispheres for orthographic encoding.

One aspect of the results of Exp. 1 that is difficult to interpret concerns the effect of lexical frequency and its interaction with priming. Under a strict interpretation of the outcome of the data analyses where only differences with $P < .05$ are considered real, IH only differs from normal observers by the fact that his reading accuracy is lower for low-frequency than high-frequency words. However, there are weaker aspects of the data that suggest more fundamental differences between IH and normal observers regarding the lexical frequency effect. It was noted that normal subjects showed a marginally significant reduction of RTs with high- relative to low-frequency words, whereas no indication for such an effect was present in IH. Furthermore, whereas it is clear that neurologically intact subjects showed a priming effect of equal magnitude with high- and low-frequency words, Fig. 4 suggests that priming in IH may have been somewhat weaker with low-frequency words. What seems to be a major reason for the failure of the RT data analysis in IH to demonstrate such an interaction is the fact that it was based on relatively few trials, given the subject's elevated error rates with low-frequency words. Another, more general reason that may have prevented the observation of clear and consistent data with respect to lexical frequency is the relatively weak manipulation of this factor in Exp. 1. Indeed, whereas low-frequency words had an occurrence frequency ranging between 3 and 20 per million, the lower bound for high-frequency words was of only of 50 occurrences per million. In Exp. 3, described below, a much larger lexical frequency discrepancy is used between low- and high-frequency words and the results regarding the effect of this factor are rather clear-cut in showing an abnormal effect of lexical frequency on covert lexical activation in IH.

Based on the word priming studies conducted so far with IH, no clear limit on the capacity of covert lexical activation has emerged. Thus, a priming effect of normal magnitude occurred with 100msec primes printed in a case-alternated format and priming occurred for both low- and high-frequency words (see also Bowers, Arguin, et al., 1996, for priming effects as a function of word frequency). Moreover, as shown in Bowers, Arguin, et al., the word priming effect is highly specific since it does not generalise to orthographic neighbours. This suggests that the 100msec masked primes used result in a very accurate activation of abstract orthographic word forms. However, provided that the covert lexical activation assumed to mediate the word priming effect bears any relation to the overt word recognition performance of the patient, as argued in the Introduction one should eventually be able to find a limit to the capacities of covert lexical activation, which would signal an abnormal functional bottleneck that may be at the origin of the LBL disorder. One such limit, suggested by Bowers, Arguin et al., is access to phonological representations of words. It is conceivable that failure of such access may be responsible for the reading deficit in LBL dyslexia, although it would not prevent fast abstract orthographic priming effects such as those noted so far in IH. Experiment 2 provides a test of this possibility by examining phonological priming.

EXPERIMENT 2

To assess the possibility that phonological access constitutes a significant limit on the covert lexical activation process in IH, Exp. 2 used a word priming procedure similar to that used in Exp. 1. The same set of targets was tested under repeated, unrelated, and homophone priming conditions. In the last condition, the prime was orthographically distinct from the target but was homophonic to it. Previous research in neurologically intact observers has shown robust benefits from primes that are homophones to the target in a word naming task (Lukatela & Turvey, 1994).

Methods

Subjects were IH and a group of 15 neurologically intact individuals (5 males, 10 females) aged between 18 and 20. As for Exp. 1, the latter subjects served to determine that the results expected from an intact mature reading system actually occurred under the experimental conditions used here. The procedure was the same as in Exp. 1, except that primes were printed in lower-case letters whereas targets were printed in upper-case letters. Targets (Appendix B) were 44 four- and five-letter words, ranging in frequency from 1 to 298 per million, which had at least three of their component letters with very different shapes between upper- and lower-case formats (a/A, b/B, d/D, e/E, g/G, l/L, m/M, q/Q, r/R). Because of the patient's difficulty in reading irregular words, these were avoided as much as possible. Only three of the targets used in Exp. 2 have irregular spelling-to-sound correspondences. Each target was tested under three priming conditions. Repeated: the prime was the same word as the target; Unrelated: the prime was a word of the same length as the target but had no orthographic overlap and was phonologically different from the target; Homophone: the prime was a word orthographically different from the target but had the same pronunciation (e.g. prime = gait; target = GATE). The complete trial list comprised 3 blocks of 44 trials with no target repeated within a block. To ensure enough observations per condition, each subject was administered the complete stimulus list twice. For IH, these repeated administrations occurred in different sessions separated by an interval of 2 weeks. The order in which blocks were run with neurologically intact subjects was counterbalanced across subjects. For IH, a total of three trials (1.1%) were lost due to the failure of the subject's response to trigger the microphone. Across all trials run with neurologically intact subjects, 17 trials (0.9%) were lost due to a microphone error. These trials were not considered in the data analysis.

Results

Average correct RTs and error rates observed in neurologically intact subjects are presented in Figs. 6 and 7, respectively. The correlation between RTs and error rates was exact and positive (+ 1.00, $P < .05$), which indicates no speed-accuracy trade-off. For each subject, RTs that were more than 3 SDs away from the mean for their condition were discarded. A total of 37 data points (1.9% of trials) were removed from the analysis on this criterion. An ANOVA carried out on correct RTs showed a significant effect of priming [$F(2,28) = 21.3; P < .001$]. RTs with Repeated and Homophone primes were both significantly lower than those with Unrelated primes [$t(28) = 5.9; P < .001; t(28) = 4.1; P < .005$; respectively]. In addition, RTs with Repeated and Homophone primes did not differ significantly $t(28) = 1.6$; n.s.]. The analysis performed on error rates showed no effect of priming [$F(2,28) < 1$]. Thus, although perfectly correlated with RTs, error rates were very low (overall average of 1.4%), and the difference between any pair of conditions did not exceed 1%.

Average correct RTs and error rates for IH are shown in Figs. 8 and 9, respectively. The correlation between RTs and error rates was of + 0.65 (n.s.), which indicates no speed-accuracy trade-off. For IH, no correct response latency was found which was more than 3 SDs away from the mean for its condition. An ANOVA carried out on correct RTs showed a significant effect of priming [$F(2,222) = 6.1; P < .01$]. RTs with Repeated primes were shorter than those with Unrelated [$t(222) = 3.5; P < .001$] or Homophone primes [$t(222) = 2.0; P < .05$]. By contrast, RTs with Homophone and Unrelated primes did not differ significantly [$t(222) = 1.5$; n.s.]. The effect of priming on error rates was not significant [$\chi^2(2) = 2.4$; n.s.].

Discussion

The results of Exp. 2 replicate the abstract word repetition priming effect previously observed in IH. The magnitude of the repetition priming effect (relative to Unrelated primes) shown by the patient in relation to overall correct RTs is slightly higher (12.9%) than that observed in neurologically intact subjects (9.5%), thus indicating that IH was at least as sensitive to the word repetition effect as were normal individuals. However, whereas neurologically intact subjects also showed a substantial RT benefit with Homophone primes relative to Unrelated primes, thereby replicating previous observations by Lukatela and Turvey (1994), the difference between these conditions was not significantly for IH.

The occurrence of a homophone priming effect in neurologically intact observers im-

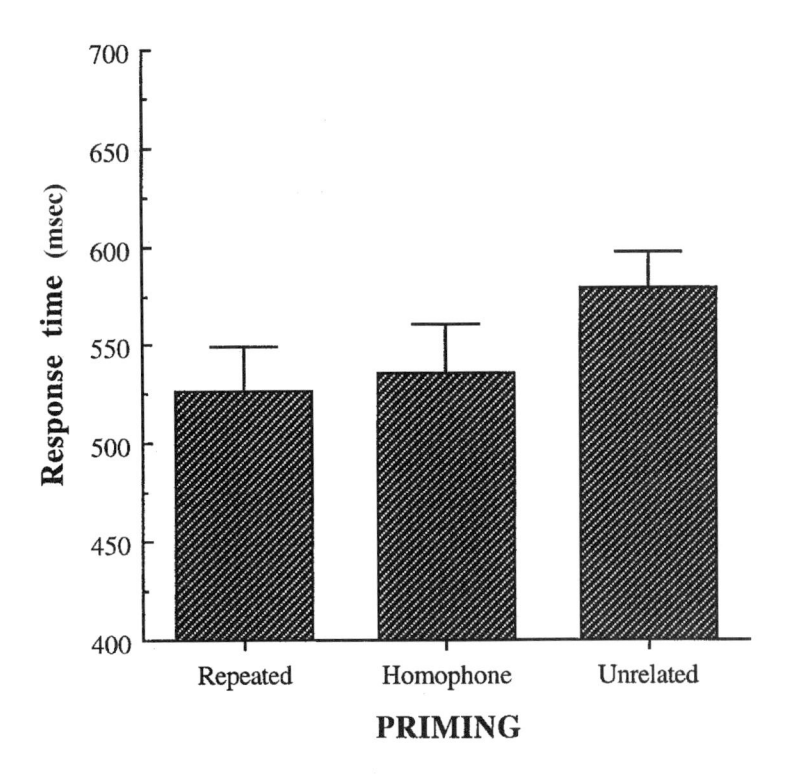

Fig. 6. Average correct RTs to targets preceded by repeated, homophone, or unrelated primes in neurologically intact subjects (Exp. 2).

plies the advance activation of the phonological representation of the target by the prime. That this effect failed to occur in IH indicates that covert lexical activation fails to reach the phonological representations of words. This limit on covert lexical activation in IH, which was initially hypothesised by Bowers, Arguin, et al. (1996), was proposed by these authors as the basic cause for LBL reading in the patient. Thus, in light of repetition priming results, which showed no clear limitation on orthographic encoding capacity in IH, it was proposed that the source of the LBL reading disorder must lie further in the processing stream involved in overt word recognition and one obvious candidate was phonological access. Thus, according to this proposal, it is a failure in the transfer of an intact global orthographic activation to phonological representations of words that would prevent normal reading performance in IH and force the patient to resort to what appears as a letter-by-letter decoding strategy. The results obtained here in Exps. 1 and 2 are largely consistent with this theory. Thus, neither experiment suggests a significant aberration in the patient's capacity for orthographic encoding, but Exp. 2 clearly argues for a failure of phonological access. What constitutes an important difficulty with this view, however, is

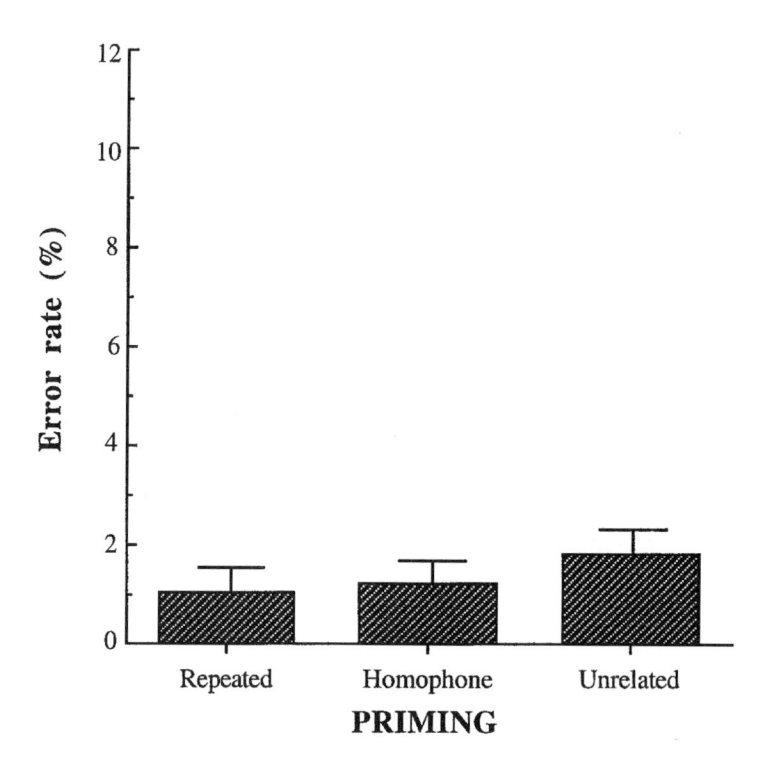

Fig. 7. Percentage error rates to targets preceded by repeated, homophone, or unrelated primes in neurologically intact subjects (Exp. 2).

that its viability is largely dependent on the absence of anomalous findings with respect to orthographic encoding; i.e. on negative results. Stronger support for the proposal of Bowers, Arguin, et al. (1996) may then require additional, and possibly more stringent tests of orthographic encoding capacity in IH.

Exp. 3 will provide a further assessment of the assumption of intact orthographic encoding processes in IH. However, the method used will be quite different from the priming paradigm we have applied so far. Thus, the task will simply consist in reading words aloud and, instead of the word priming effect, the index for covert lexical activation as modulating overt recognition performance will be based on the facilitatory effect of increased orthographic neighbourhood size.

EXPERIMENT 3

Previous studies of visual word recognition in neurologically intact observers have shown that the orthographic similarity of a target with other words of the vocabulary affects the time required to recognise it. In particular, it has been shown that targets with many orthographic neighbours (i.e. other words of the same length that differ from it by just one

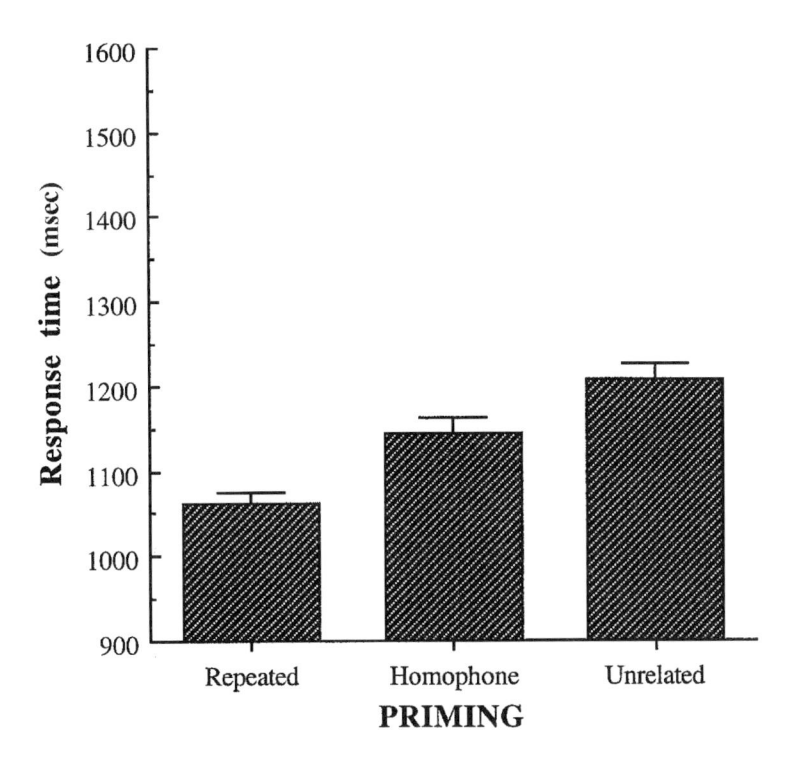

[Graph showing response time (msec) on y-axis from 900 to 1600, with three bars for priming conditions on x-axis: Repeated (~1060), Homophone (~1150), Unrelated (~1210).]

Fig. 8. Average correct RTs to targets preceded by repeated, homophone, or unrelated primes in IH (Exp. 2).

letter) may be recognised faster than words with few orthographic neighbours (Andrews, 1989, 1992; Forster & Shen, 1996; Peereman & Content, 1995; Sears, Hino, & Lupker, 1995). Although the exact cause for this facilitatory effect of increased orthographic neighbourhood size is still unclear (Forster & Shea, 1996), its occurrence is generally assumed to depend on a global (i.e. parallel) activation of orthographic word forms (Andrews, 1992; Peereman & Content, 1995; Sears et al., 1995).

In contrast, if overt word recognition was conducted by a strictly letter-by-letter procedure, as often assumed for LBL readers, one should expect the effect of increased orthographic neighbourhood size to be inhibitory. Assume, for instance, a simple word recognition model in which a letter processing module sequentially feeds information about letter identities to another module representing the orthographic forms of words. Assume also, as suggested by observations by Arguin and Bub (1995; see Luce, 1959, 1977, for a detailed discussion), that overt recognition of the target is achieved once the ratio of activation of its lexical representation (i.e. signal) over the activation of other lexical representations (i.e. noise) exceeds some fixed threshold. With every letter identity that is sequentially passed to the word-form system, the activation of the target and of any other word compatible with the

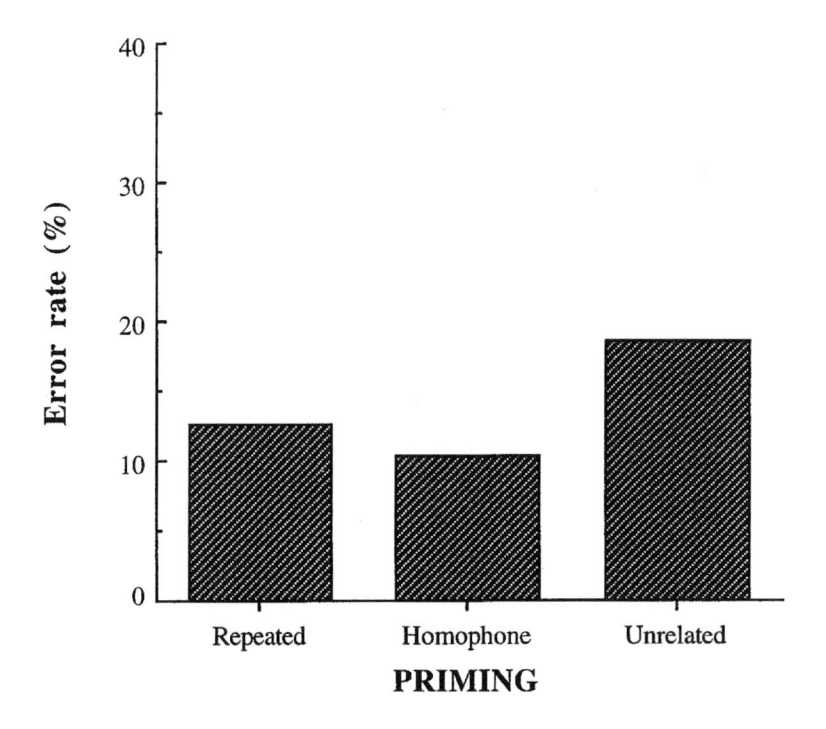

Fig. 9. Percentage error rates to targets preceded by repeated, homophone, or unrelated primes in IH (Exp. 2).

letter input received up to that point will increase to the same degree (assuming everything else is equal). Only once an incompatible letter identity is encountered will the activation of a nontarget representation begin to be lower than that of the target, and presumably this activation should decay over a period of time rather than vanish immediately. Statistically, what this means is that, with serial letter input, nontarget representations should be activated in greater numbers, to a greater degree, and for a longer duration if the target has many orthographic neighbours than if it has few or none. This increased background noise against which the activation of a target word with many orthographic neighbours must be assessed should be costly in terms of overt recognition performance. By contrast, if the letter input to the word form system is parallel, letter information incompatible with orthographic neighbours of the prime is received at the same time as compatible letter identities. This should keep the activation of orthographic neighbours of the prime sufficiently low from the outset that any noise they produce within the lexical system remains manageable and does not prevent whatever facilitatory effect these neighbours may otherwise have on target processing to be manifest in performance. Congruent with the notion that orthographic neighbours may negatively affect reading performance when incomplete letter identity information is passed to the word-form system—as it is for some duration if reading is strictly letter-by-letter—are observations from a patient with neglect dyslexia (Arguin & Bub, 1997). This patient very often tended to ignore the first letter of words in her reading attempts and her results suggest that orthographic neighbours of the target that differed from it on their first letter were strongly activated. Thus, when the target had many such neighbours that were of a higher frequency than itself, the patient's neglect error rate was doubled relative to when the target had no such neighbours.

According to the view presented here, if overt word recognition in LBL reading is exclusively mediated by sequential letter identification, performance should be negatively affected by an increase in orthographic neighbourhood size. In contrast, the observation of a facilitatory effect of increased orthographic neighbourhood size in an LBL reader would argue for a contribution of covert orthographic lexical activation (i.e. lexical access based on parallel letter processing) to overt word recognition performance. The central aim of Exp. 3 is to provide a test of these contrasting predictions.

Another factor that has been shown to affect word recognition times in neurologically intact subjects, and which may be assumed to result at least in part from the activation of the global orthographic form of the target, is lexical frequency (e.g. Monsell, Doyle, & Haggard, 1989; Paap, McDonald, Schvaneveldt, & Noel, 1987; Waters & Seidenberg, 1985). In another paper of this issue, the literature review of Behrmann et al. (this issue) shows that LBL readers also generally show benefits from increased lexical frequency, and the authors suggest that this benefit must result from a parallel input to the word-form system since it increases as a function of word length. In the present experiment, the word-frequency effect will be used not so much as a direct index for the occurrence of covert lexical activation, but rather as a way to characterise this covert activation by examining how the frequency effect interacts with orthographic neighbourhood size. Indeed, in neurologically intact observers, the facilitatory effect of increased neighbourhood size is greater with, or exclusive to, low-frequency words (Andrews, 1989, 1992; Peereman & Content, 1995; Sears et al., 1995). Inasmuch as a facilitatory effect of increased orthographic neighbourhood size may be found in an LBL reader, the interaction of this factor with lexical frequency should help characterise the mechanisms responsible for covert lexical activation.

The effect of orthographic neighbourhood size and its interaction with lexical frequency were examined here by having IH and a group of neurologically intact subjects read aloud a series of four-letter words that varied orthogonally on their numbers of orthographic neighbours and their lexical frequencies.

Methods

Subjects were IH and a group of 15 neurologically intact individuals (5 males, 10 females) aged between 18 and 20. Normal subjects served to determine that the results expected from an intact reading system actually occur under the experimental conditions used here. Each trial began with a 1500msec fixation point, displayed at the centre of the computer screen, on which subjects were instructed to keep their eyes fixated. This was followed by an upper-case word target whose right extremity was aligned 1cm (1.1°; from a viewing distance of about 50cm) to the left of fixation. The target was printed in upper case and remained visible until response, which was then typed in by the experimenter before the next trial was initiated by a keypress. The subject's task was to name the target as rapidly as possible while avoiding errors. As in previous experiments, response times were registered by a voice-key connected to the computer controlling the experiment. Target words (N = 50 per condition; Appendix C) varied orthogonally on their numbers of orthographic neighbours (Low range: 0–3; High range: 11 or more) and their lexical frequencies (Low range: 1–15; High range: 100 or more). Across conditions, words were matched quadruplet-wise on single-letter (average sum across words = 833.8) and bigram frequencies (average sum across words = 244.7). As in the previous experiments, words with irregular spelling-to-sound correspondences were avoided because the patient is more likely to commit naming errors with such words. Throughout the list there was a total of eight irregular words; four were low frequency/low neighbourhood size, one low frequency/high neighbourhood, one high frequency/low neighbourhood and two high frequency/high neighbourhood. To ensure enough observations per condition, the complete stimulus list was administered to IH twice in different sessions separated by an interval of 2 weeks. For IH, a total of 11 trials (2.8%) were lost due to the failure of the subject's response to trigger the microphone. Across all trials run with neurologically intact subjects, 18 trials (0.6%) were lost due to a microphone error. These trials were not considered in the data analysis.

Results

Average correct RTs and error rates for normal subjects are presented in Figs. 10 and 11, respectively. The correlation between RTs and error rates was high and positive (+ .98; $P < .05$), which indicates no speed-accuracy trade-off. For each subject, RTs that were more than 3 SDs away from the mean for their condition were discarded. A total of 18 data points (0.6% of trials) were removed from the analysis on this criterion. An ANOVA conducted on correct RTs with Orthographic neighbourhood size and Lexical frequency as factors showed main effects of neighbourhood size [$F(1,14) = 10.6$; $P < .01$] and of frequency [$F(1,14) = 16.2$; $P < .005$], as well as a marginally significant interaction [$F(1,14) = 3.9$; $P < .07$]. The main effects indicated shorter RTs to words with many orthographic neighbours and to high-frequency words. Simple effects of neighbourhood size as a function of frequency showed a significant effect of number of neighbours with low-frequency items [$t(14) = 3.1$; $P < .01$], but none with high-frequency words [$t(14) = 1.3$; n.s.]. The outcome of the analysis of error rates paralleled that with RTs. Thus, main effects of neighbourhood size [$F(1,14) = 16.0$; $P < .005$]

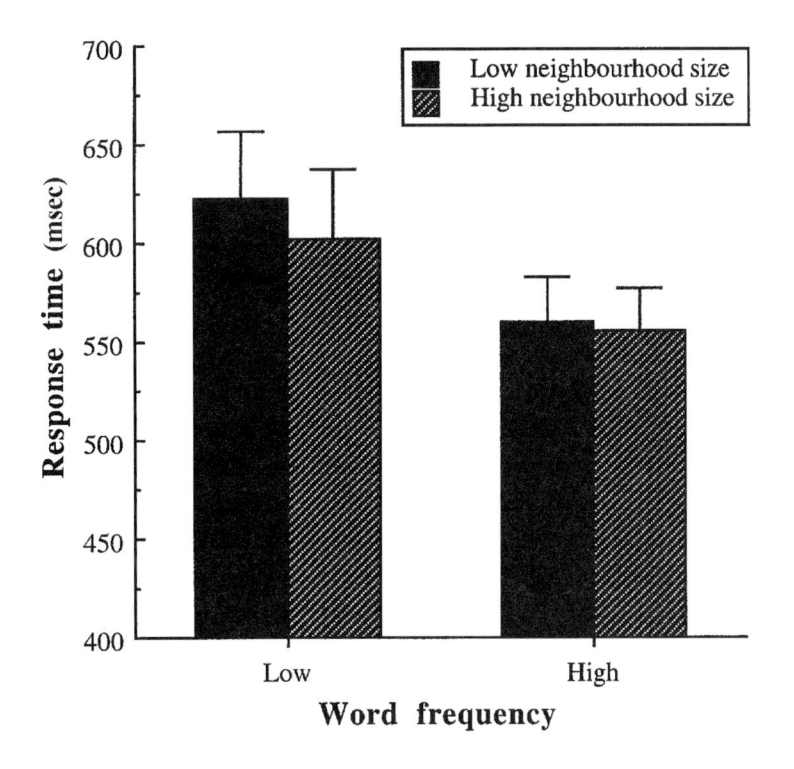

Fig. 10. Average correct RTs as a function of orthographic neighbourhood size and frequency of the target in neurologically intact subjects (Exp. 3).

and frequency [$F(1,14) = 22.1$; $P < .001$] were present, as well as a significant neighbourhood size × frequency interaction [$F(1,14) = 6.8$; $P < .05$]. Error rates were reduced with high neighbourhood size targets and with words that had a high frequency. Simple effects of the interaction showed a significant effect of neighbourhood size with low frequency words [$t(14) = 3.7$; $P < .005$] but not with high-frequency words [$t(14) = 1.2$; $n.s.$].

Average correct RTs and error rates for IH are shown in Fig. 12 and 13, respectively. The correlation between RTs and error rates was of + .95 ($P = .05$), which indicates no speed-accu-racy trade-off. Response latencies that were more than 3 SDs away from the mean for their condition were discarded from the analysis of correct RTs. Seven data points were removed from the analysis on this criterion. A two-way ANOVA of Orthographic neighbourhood size × Lexical frequency showed significant main effects of neighbourhood size [$F(1,270) = 58.9$; $P < .001$] and of frequency [$F(1,270) = 9.1$; $P < .005$], but no interaction [$F(1,270) < 1$]. The main effects indicate shorter RTs with words with a large orthographic neighbourhood size and with high-frequency targets. Analysis of error rates showed a significant reduction of error rates with increased orthographic neighbourhood size [$\chi^2(1) = 16.4$; $P < .001$] but no effect of lexical frequency [$\chi^2(1) = 0.1$; $n.s$].

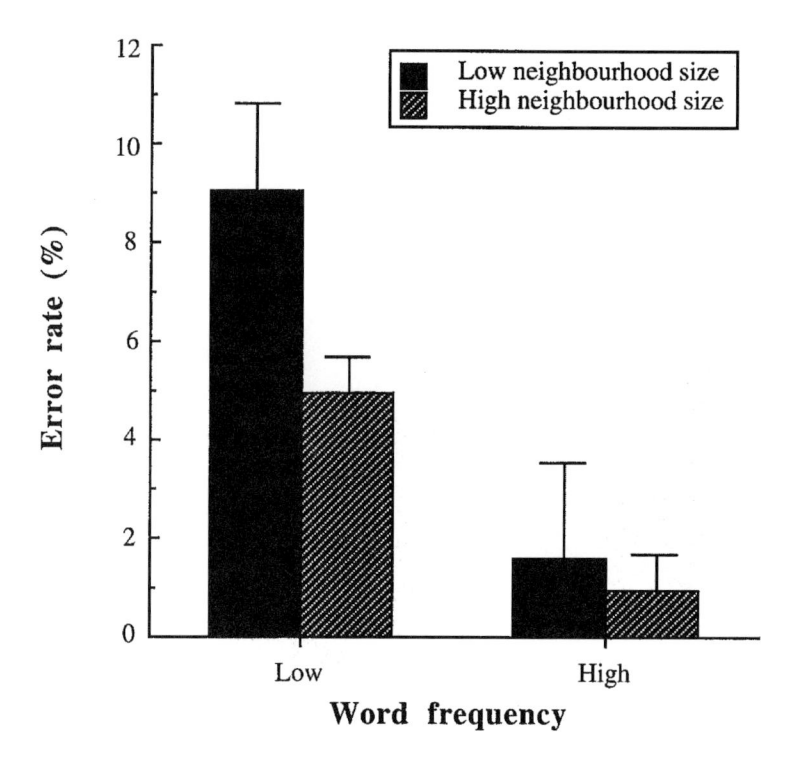

Fig. 11. Percentage error rates as a function of orthographic neighbourhood size and frequency of the target in neurologically intact subjects (Exp. 3).

Discussion

The results of Exp. 3 have shown a very substantial facilitation of overt word recognition performance in IH by an increase in the number of orthographic neighbours the target has. Thus, increased neighbourhood size led to an overall 222msec reduction in RTs and to half as many errors as with low neighbourhood size targets (Figs. 5 and 6). As discussed in the introductory section of this experiment, this effect is incongruent with the hypothesis of overt word recognition in LBL reading being mediated strictly by an LBL process. Rather, these observations point to an important contribution of covert orthographic lexical activation to overt word recognition performance. This is congruent with the word priming results previously observed in IH, whereby the prior activation of the orthographic representation of the target by the prime facilitated its overt recognition.

In contrast to priming results from IH, however, one interesting aspect of Exp. 3 is that it suggests a major discrepancy between the patient and neurologically intact observers on the activation of the orthographic forms of words. Thus, whereas IH showed significant and equal facilitation from increased orthographic neighbourhood size with low- and high-frequency words, the effect occurred only with

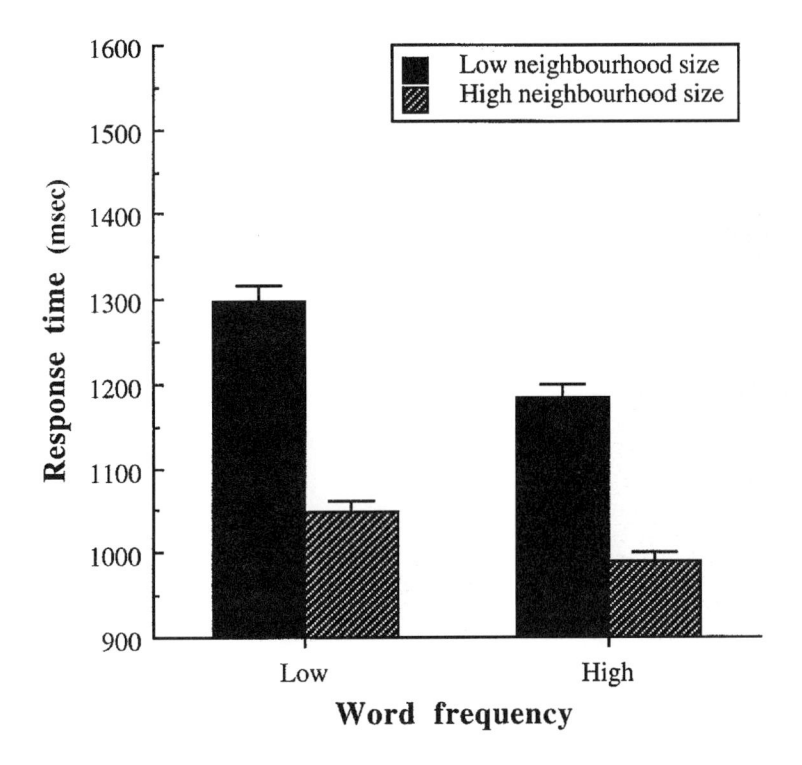

Fig. 12. Average correct RTs as a function of orthographic neighbourhood size and frequency of the target in IH (Exp. 3).

low-frequency words in normal subjects. The form of the interaction observed in normals suggests that although both increased neighbourhood size and frequency facilitate word recognition, this facilitation saturates when these two factors are combined. That is, with high-frequency words, it seems that there was no room for further performance improvement with increased neighbourhood size. No such saturation is apparent in IH's results, however, because increased orthographic neighbourhood size facilitated performance just as much with low- and high-frequency words. This lack of an interaction in IH is not simply due to insensitivity of the patient to lexical frequency, as he showed a substantial reduction of response latencies with high-frequency words[3]. Rather, it appears that, in contrast to normals, even with words of high frequency and large neighbourhood size, the activation of orthographic word forms in IH is still not optimal and that there is room for it to be improved further. This points to an anomaly in the activation of lexical representations in IH, and in particular it suggests that this

---

[3] Note that this effect cannot be attributed to the frequencies of sublexical components of words because items were matched on single-letter and bigram frequencies.

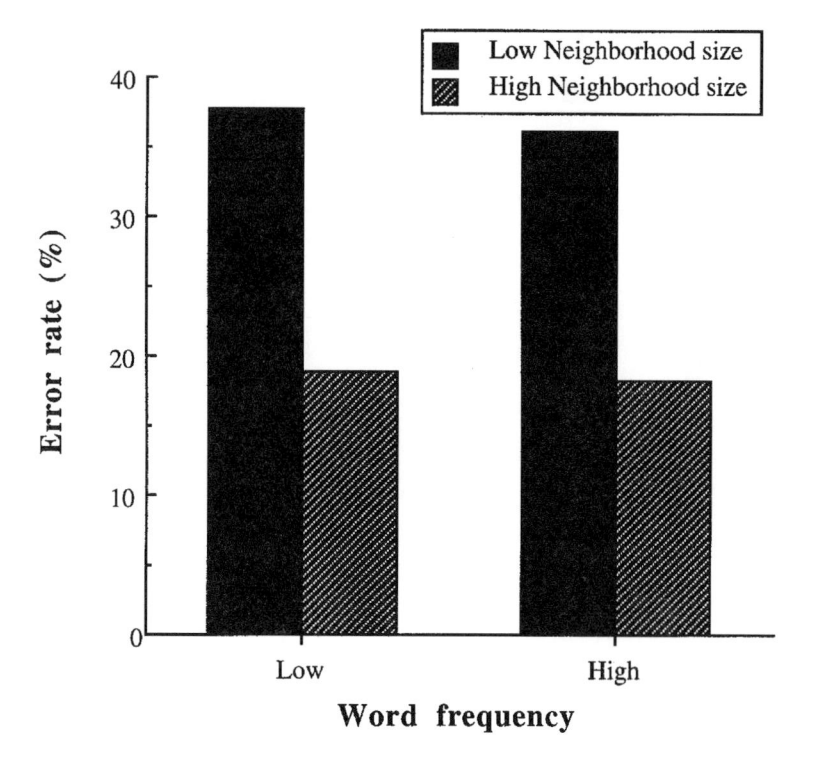

Fig. 13. Percentage error rates as a function of orthographic neighbourhood size and frequency of the target in IH (Exp. 3).

activation may be weaker than in normal readers, thus explaining the absence of a saturation effect in the patient's results in Exp. 3. This characterisation of covert lexical activation in IH implies that it may be a limitation on the activation of orthographic word forms, achieved through a parallel processing of letters, which forces the patient to rely on a serial letter identification strategy for the overt recognition of words. This view, which is similar to that proposed by other investigators to explain implicit reading in their LBL patients, is discussed in greater detail in the next section.

GENERAL DISCUSSION

The phenomenon of implicit reading in brain-damaged patients suffering from letter-by-letter reading suggests a process of covert lexical activation, whereby some form of lexical access—which fails to support overt word recognition reliably—occurs rapidly on the basis of the parallel processing of the constituent letters of the stimulus. An accurate characterisation of covert lexical activation in LBL readers through the use of implicit reading evidence has proven difficult because several patients do not show the phenomenon, which also appears sensitive to strategy effects. However, previous observations (Bowers, Arguin, et al., 1996; Bowers, Bub, et al., 1996; Howard,

1991) suggested that covert lexical activation may affect overt word recognition performance in LBL readers. On this basis, the present paper aimed to specify the extent as well as some limits of covert lexical activation in LBL dyslexia by studying the effect of variables assumed to denote such activation on the overt word recognition performance of patient IH. The variables studied here were those of masked abstract orthographic (Exp. 1) and homophone (Exp. 2) priming, as well as the effect of orthographic neighbourhood size of a target and the interaction of this factor with lexical frequency (Exp. 3).

In Exp. 1, masked identity primes printed in a case-alternated format and shown for 100msec substantially reduced IH's reading latency for upper-case words relative to unrelated primes. In addition, the magnitude of the repetition priming effect expressed in relation to overall average RTs was quite similar to that found in neurologically intact observers in that same experiment. These observations suggest that the covert lexical activation process mediating the priming effect is based on a truly abstract orthographic encoding mechanism similar to that characterising normal reading. This result also suggests that covert lexical activation may depend on the residual function of a left-hemisphere reading mechanism since it appears, according to our current knowledge, that the right cerebral hemisphere may not be capable of supporting priming under the conditions of Exp. 1. One clear limitation of covert lexical activation in IH was demonstrated in Exp. 2 by a failure of primes that were homophones to the target to affect the patient's overt word recognition performance significantly, in contrast to observations from neurologically intact observers. This suggests that no significant degree of covert phonological activation occurs in IH. In Exp. 3, results from IH as well as from normal readers showed a facilitatory effect of increased orthographic neighbourhood size on target identification, and it was argued that such an effect could not possibly occur if reading was effected exclusively through a LBL process. The covert orthographic lexical activation implied by the neighbourhood size effect in IH appears abnormal, however. Indeed, although this effect was equally strong with high- and low-frequency words in the patient, it occurred only with low-frequency words in our normal readers. It was proposed that IH's results in that experiment may best be explained by the hypothesis that only weak activation of orthographic lexical representations is achieved through a parallel processing of letters in IH, and that this may be the reason why this form of lexical activation cannot reliably support overt word recognition.

Accounts similar to that proposed here for covert lexical activation have been offered before by a number of different authors to explain evidence for implicit reading in LBL patients (Arguin & Bub, 1993; Bub & Arguin, 1995; Bub et al., 1989; Shallice & Saffran, 1986). Essentially, the notion proposed is that the lexical activation is achieved rapidly and by a parallel analysis of letters, but that this process fails to provide an activation contrast between the target and other words that is sufficient for overt recognition. The sheer presence of some degree of activity among lexical representations may be sufficient, however, to perform classification tasks such as lexical or semantic decisions. Indeed, it may be assumed that such tasks are less demanding than that of overt recognition with respect to the quality required for the internal representation of the stimulus to maintain an accurate performance. Similarly, the presence of only a weak degree of lexical activation may be sufficient to facilitate the identification of the constituent letters of words relative to those of nonwords. Exactly the same reasoning may hold with respect to covert lexical activation effects on overt word identification. Thus, even weak lexical activation may be sufficient for word recognition to be facilitated by masked priming, for instance. Also, if an increase in orthographic neighbourhood size is assumed to facilitate the activation of the target representation obtained by a parallel encoding of the letters, either directly or via a feedback facilitation of letter processing, even weak covert lexical activation may account for the neighbourhood size effect in IH.

Clearly, however, not all codes that serve for the internal representation of words may be addressed equally well by covert lexical activation. With IH, for instance, orthographic activation, although apparently weak, was sufficient to sustain abstract priming between a lower-case or a case-alternated prime and an upper-case target, as well as the orthographic neighbourhood size effect. However, in the same patient, no significant evidence for covert phonological activation could be found in the homophone priming condition. This could have little implication for semantic processing though, so a test of covert semantic activation in IH might or might not have revealed such an effect.

When proposed, the notion of weak covert lexical activation to account for implicit reading in LBL patients is often accompanied by the assumption that representations of words similar to the target may be activated to an excessive degree or that such items are insufficiently suppressed (e.g. Arguin & Bub, 1993; Shallice & Saffran, 1986). This implies that the selectivity of covert lexical activation is deficient and that overt recognition is prevented not only by the poor signal provided by the activation of the target, but also by the high degree of noise caused by the excessive activation of other words. It seems that this may not be the case for IH, however, since masked priming does not generalise to orthographic neighbours of the target (Bowers, Arguin, et al., 1996). This observation argues for the preserved selectivity of covert lexical activation in the patient.

One important implication of the findings reported here concerns the type of processing by which overt word recognition is performed in LBL readers (Hanley & Kay, 1992; Howard, 1991; Rapcsak, Rubens, & Laguna; 1990; Warrington & Shallice, 1980). Since the discovery by Déjerine (1891) that left occipital damage may cause LBL reading, it has largely been assumed that overt word recognition in these patients is exclusively mediated by the sequential recognition of individual letters, mainly on the basis that several patients overtly name individual letters before being able to recognise a target word. In more recent years, the principal evidence that has been invoked in support for the claim that reading is essentially based on sequential letter identification is that reading times in LBL patients increase linearly with the number of letters in the target. This linear effect of number of letters does indeed strongly suggest that individual letters are processed serially in the overt word-recognition performance of LBL readers. However, it appears that this is not the only mode of lexical access for LBL readers.

One indication of this was provided by the observation of implicit reading in LBL patients, which implied a lexical access based on the parallel processing of letters. This evidence, however, did not tell us whether this form of lexical access actually contributed to overt word recognition performance. Such an indication was provided later by Howard (1991), who showed that some proportion of overt word reading responses was indeed based on parallel letter processing. Again, however, the results did not indicate whether parallel lexical access had any contribution to overt reading performance when patients resorted to an LBL process for word recognition. More recently, the word priming observations of Bowers, Arguin, et al. (1996), as well as those provided in Exps. 1 and 2 of the present paper, went somewhat further in showing that parallel letter processing could significantly contribute to overt word recognition and that this contribution was not restricted to a small subset of "anomalous" trials such as those studied by Howard (1991), but rather that it occurred consistently. However, in those experiments, evidence for parallel letter encoding essentially referred to the processing of the prime, not that of the target. It thus remained possible that overt recognition of the target itself was strictly based on a serial LBL process, even through this recognition performance was facilitated by the prior parallel processing of the prime. The facilitatory neighbourhood size effect (reported in Exp. 3), though, strongly suggests that parallel letter encoding (i.e. covert lexical activation) provides a direct and consistent contribution to the overt recognition of a word. This experiment did not assess the word length effect in IH to determine that his overt recognition performance effectively involved a sequential processing of letters. However, such serial processing in IH has been documented on several previous occasions spread across a period of 5 years, during which the magnitude of the word-length effect has remained essentially unchanged. It would seem reasonable, then, to assume that overt recognition of the targets by IH in Exp. 3 involved the serial processing of individual letters. What the facilitatory effect of increased orthographic neighbourhood size in IH suggests, therefore, is that parallel and serial letter processing mechanisms may provide a conjoint contribution to overt word recognition performance in LBL reading.

Interestingly, on the basis of distinct indicators that are also different from those used in the present paper, reports by Behrmann et al. (this issue) and by Montant et al. (this issue) also argue for conjoint effects of parallel and serial letter processing on the overt word recognition performance of LBL readers. As discussed in the introduction, this conjoint contribution may result from separate serial and parallel letter processing mechanisms that converge onto a common lexical system that is directly responsible for overt word recognition. Alternatively, one may assume the operation of a single reading system in which lexical activation from the parallel processing of letters is possible, but this activation must be supplemented by the focused processing of individuals letters to render overt word recognition possible. Behrmann et al. and Montant et al. have opted for the second, more parsimonious account, since they found no indication suggesting the need to assume separate systems mediating the parallel and serial processes. However, previous observations indicating a puzzling dissociation between word and letter priming in IH suggest the possibility of separate systems mediating covert lexical activation and serial letter identification for overt word recognition.

Bowers, Arguin et al. (1996), using a priming procedure identical to that of Exp. 1 but in a task of single-letter identification, have shown a deficit in abstract letter encoding in IH (see also Arguin & Bub, 1994; for similar findings in LBL patient DM). Thus, a masked prime that is nominally identical to the target but visually different from it (e.g. a/A) had no effect relative to an unrelated prime with a prime duration of 100msec, even though physical identity priming caused large benefits. This is markedly different from what was previously observed by Arguin and Bub (1995) in neurologically intact subjects. They showed very substantial benefits with nominally identical primes with a prime duration as short of 100msec, and these benefits did not differ from those obtained with physically identical primes. The absence of abstract letter priming in IH with 100msec primes is in striking contrast to his performance in the word priming task, which shows large benefits from abstract repetition priming with a prime duration of 100msec. This qualitative dissociation between letter and word priming in IH suggests that two separate reading mechanisms may be active in the patient. One of them, responsible for covert lexical activation effects, would be abstract with respect to visual shape. The other, serving for the overt identification of isolated letters, would be shape specific. If this latter mechanism is also responsible for the sequential processing of strings of letters in overt word recognition tasks, it would mean that the parallel and serial processes involved in such tasks are mediated by separate systems.

Such a possibility raises another crucial issue concerning efforts directed to a specification of the functional impairment(s) responsible for LBL reading. The logic commonly employed for this purpose is to attribute the high-level word recognition disorder to some demonstrated impairment of a low-level process on which it depends for normal performance, even if it is sometimes difficult to provide a clear and detailed functional account of the relation between cause and effect. In the case of IH, for instance, this logic could attribute the reading disorder to the deficit in abstract orthographic encoding suggested by the letter priming results. It seems this account may be mistaken, however, given the qualitative dissociation between word and letter priming that is shown by the patient. Rather, it appears that the letter priming results observed in IH may be relevant not so much to specify the cause of his reading disorder, but rather to characterise some compensatory process the patient must rely on for overt word recognition. What this means, then, is that caution should be exercised in assigning a causal relationship between LBL reading and other concomitant impairments, since these associated deficits may in fact reflect a form of adaptation to the reading disorder rather than its cause.

CONCLUSIONS

The investigation of covert lexical activation in LBL patient IH has shown that this process is based on an abstract orthographic encoding mechanism comparable to that mediating reading in neurologically intact observers, and that this process may depend on the residual function of the damaged left hemisphere. Two anomalies of covert lexical activation in IH were identified, however: (1) it does not extend to the activation of the phonological representations of words; and (2) orthographic activation resulting from a parallel encoding of letters may be particularly weak compared to that achieved in normal readers. The latter may be fundamental for the obligation of LBL patients to resort to serial letter processing for overt word recognition. Finally, results suggest that serial and parallel letter processing mechanisms contribute conjointly to the patient's overt word recognition performance.

REFERENCES

Andrews, S. (1989). Frequency and neighbourhood effects on lexical access: Activation or search? *Journal of Experimental Psychology: Learning, Memory, and Cognition, 15*, 802–814.

Andrews, S. (1992). Frequency and neighbourhood effects on lexical access: Lexical similarity or orthographic redundancy? *Journal of Experimental Psychology: Learning, Memory, and Cognition, 18*, 234–254.

Arguin, M., & Bub, D.N. (1993). Single-character processing in a case of pure alexia. *Neuropsychologia, 31*, 435–458.

Arguin, M., & Bub, D. (1994). Pure alexia: Attempted rehabilitation and its implications for interpretation of the deficit. *Brain and Language, 47*, 233–268.

Arguin, M., & Bub, D. (1995). Priming and response selection processes in letter classification and identification tasks. *Journal of Experimental Psychology: Human Perception and Performance, 21*, 1199–1219.

Arguin, M., & Bub, D. (1997). Lexical constraints on reading accuracy in neglect dyslexia. *Cognitive Neuropsychology, 14*, 765–800.

Behrmann, M., Black, S.E., & Bub, D. (1990). The evolution of pure alexia: A longitudinal study of recovery. *Brain and Language, 39*, 405–427.

Behrmann, M., Plaut, D.C., & Nelson, J. (this issue). A literature review and new data supporting an interactive account of letter-by-letter reading. *Cognitive Neuropsychology, 15*(1/2).

Behrmann, M., & Shallice, T. (1995). Pure alexia: A nonspatial visual disorder affecting letter activation. *Cognitive Neuropsychology, 12*, 409–454.

Boles, D.B. (1992). Fast visual generation: Its nature and chronometrics. *Perception and Psychophysics, 51*, 239–246.

Boles, D.B., & Clifford, J.E. (1989). An upper- and lower-case alphabetic similarity matrix, with derived generation similarity values. *Behaviour Research Methods, Instruments, and Computers, 21*, 579–586.

Boles, D.B., & Eveland, D.C. (1983). Visual and phonetic codes and the process of generation in letter matching. *Journal of Experimental Psychology: Human Perception and Performance, 9,* 657–674.

Bowers, J.S., Arguin, M., & Bub, D. (1996). Fast and specific access to orthographic knowledge in a case of letter-by-letter surface alexia. *Cognitive Neuropsychology, 13,* 525–567.

Bowers, J.S., Bub, D., & Arguin, M. (1996). A characterization of the word superiority effect in pure alexia. *Cognitive Neuropsychology, 13,* 415–441.

Bub, D.N., & Arguin, M. (1995). Visual word activation in pure alexia. *Brain and Language, 49,* 77–103.

Bub, D.N., Black, S., & Howell, J. (1989). Word recognition and orthographic context effects in a letter-by-letter reader. *Brain and Language, 36,* 357–376.

Buxbaum, L., & Coslett, H.B. (1996). Deep dyslexic phenomenon in a letter-by-letter reader. *Brain and Language, 54,* 136–167.

Coltheart, M., Davelaar, E., Jonasson, J.T., & Besner, D. (1977). Access to the internal lexicon. In S. Dornic (Ed.), *Attention and performance VI* (pp. 535–555). London: Academic Press.

Coslett, H.B., & Monsul, N. (1994). Reading with the right hemisphere: Evidence from transcranial magnetic stimulation. *Brain and Language, 46,* 198–211.

Coslett, H.B., & Saffran, E.M. (1989). Evidence for preserved reading in "pure alexia". *Brain, 112,* 327–359.

Coslett, H.B., Saffran, E.M., Greenbaum, S., & Schwartz, H. (1993). Reading in pure alexia: The effect of strategy. *Brain, 116,* 21–37.

Damasio, A.R., & Damasio, H. (1983). The anatomic basis of pure alexia. *Neurology, 33,* 1573–1583.

Déjerine, J. (1891). Sur un cas de cécité verbale avec agraphie suivi d'autopsie. *Mémoires de la Societe Biologique, 3,* 197–201.

Farah, M.J., & Wallace, M.A. (1991). Pure alexia as a visual impairment: A reconsideration. *Cognitive Neuropsychology, 8,* 313–334.

Forster, K.I., & Guess, K. (1996). Effects of prime duration and visual degradation in masked priming. *Abstracts of the Psychonomic Society, 1,* 72.

Forster, K.I., & Shen, D. (1996). No enemies in the neighbourhood: Absence of inhibitory neighbourhood effects in lexical decision and semantic categorization. *Journal of Experimental Psychology: Learning, Memory, and Cognition, 22,* 696–713.

Friedman, R.B., & Alexander, M.P. (1984). Pictures, images, and pure alexia: A case study. *Cognitive Neuropsychology, 1,* 9–23.

Friedman, R.B., & Hadley, J.A. (1992). Letter-by-letter surface alexia. *Cognitive Neuropsychology, 9,* 185–208.

Hanley, J.R., & Kay, J. (1992). Does letter-by-letter reading involve the spelling system? *Neuropsychologia, 30,* 237–256.

Henderson, L. (1982). *Orthography and word recognition in reading.* London: Academic Press.

Howard, D. (1991). Letter-by-letter readers: Evidence for parallel processing. In D. Besner & G.W. Humphreys (Eds.), *Basic processes in reading: Visual Word Recognition* (pp. 34–76). Hillsdale, NJ: Lawrence Erlbaum Associates Inc.

Johnston, J.C. (1978). A test of the sophisticated guessing theory of word perception. *Cognitive Psychology, 10,* 123–153.

Kay, J., & Hanley, R. (1991). Simultaneous form perception and serial letter recognition in a case of letter-by-letter reading. *Cognitive Neuropsychology, 8,* 249–273.

Kinsbourne, M., & Warrington, E.K. (1962). A disorder of simultaneous form perception. *Brain, 85,* 461–486.

Kucera, M., & Francis, W. (1967). *Computational analysis of present-day American English.* Providence, RI: Brown University Press.

Levine, D.M., & Calvanio, R.A. (1978). A study of the visual defect in verbal alexia-simultanagnosia. *Brain, 101,* 65 81.

Luce, R.D. (1959). *Individual choice behaviour.* New York: Wiley.

Luce, R.D. (1977). The choice axiom after twenty years. *Journal of Mathematical Psychology, 15,* 215–233.

Lukatela, G., & Turvey, M.T. (1994). Visual lexical access is initially phonological: 2. Evidence from phonological priming by homophones and pseudohomophones. *Journal of Experimental Psychology: General, 123,* 331–353.

Marsolek, C.J., Kosslyn, S.M., & Squire, L.R. (1992). Form-specific visual priming in the right cerebral hemisphere. *Journal of Experimental Psychology: Learning, Memory, and Cognition, 18*, 492–508.

Marsolek, C.J., Squire, L.R., Kosslyn, S.M., & Lulenski, M.E. (1994). Form-specific explicit and implicit memory in the right cerebral hemisphere. *Neuropsychology, 8*, 588–597.

Mayzner, M.S., & Tresselt, M.E. (1965). Tables of single-letter and digram frequency counts for various word-lengths and letter-position combinations. *Psychonomic Monograph Supplement, 1*, 13–32.

McClelland, J.L., & Johnston, J.C. (1977). The role of familar units in perception of words and nonwords. *Perception and Psychophysics, 22*, 249–261.

Monsell, S., Doyle, M.C., & Haggard, P.N. (1989). Effects of frequency on visual word recognition tasks: Where are they? *Journal of Experimental Psychology: General, 118*, 43–71.

Montant, M., Nazir, T.A., & Poncet, M. (this issue). Pure alexia and the viewing position effect in printed words. *Cognitive Neuropsychology, 15*(1/2).

Paap, K.R., McDonald, J.E., Schvaneveldt, R.W., & Noel, R.W. (1987). Frequency and pronounceability in visually presented naming and lexical decision tasks. In M. Coltheart (Ed.), *Attention and performance XII* (pp. 221–243). Hove, UK: Lawrence Erlbaum Associates Ltd.

Patterson, K., & Kay, J. (1982). Letter-by-letter reading: Psychological descriptions of a neurological syndrome. *Quarterly Journal of Experimental Psychology, 34A*, 411–441.

Peereman, R., & Content, A. (1995). Neighbourhood size effect in naming: Lexical activation or sublexical correspondences? *Journal of Experimental Psychology: Learning, Memory, and Cognition, 21*, 409–421.

Price, C.J., & Humphreys, G.W. (1992). Letter-by-letter reading? Functional deficits and compensatory strategies. *Cognitive Neuropsychology, 9*, 427–457.

Price, C.J., & Humphreys, G.W. (1995). Contrasting effects of letter-spacing in alexia: Further evidence that different strategies generate word length effects in reading. *Quarterly Journal of Experimental Psychology, 48A*, 573–597.

Rapcsak, S.Z., Rubens, A.B., & Laguna, J.F. (1990). From letters to words: Procedures for word recognition in letter-by-letter reading. *Brain and Language, 38*, 504–514.

Rapp, B.C., & Caramazza, A. (1991). Spatially determined deficits in letter and word processing. *Cognitive Neuropsychology, 8*, 275–311.

Reicher, G.M. (1969). Perceptual recognition as a function of meaningfulness of stimulus material. *Journal of Experimental Psychology, 81*, 274–280.

Reuter-Lorenz, P.A., & Baynes, K. (1992). Modes of lexical access in the callosotomized brain. *Journal of Cognitive Neuroscience, 4*, 155–164.

Reuter-Lorenz, P.A., & Brunn, J.L. (1990). A prelexical basis for letter-by-letter reading: A case study. *Cognitive Neuropsychology, 7*, 1–20.

Saffran, E.M. (1980). Reading in deep dyslexia is not ideographic. *Neuropsychologia, 18*, 219–223.

Saffran, E.M., & Coslett, H.B. (this issue). Implicit vs. letter-by-letter reading in pure alexia: A tale of two systems. *Cognitive Neuropsychology, 15*(1/2).

Saffran, E.M., & Marin, O.S.M. (1977). Reading without phonology: Evidence from aphasia. *Quarterly Journal of Experimental Psychology, 29*, 515–525.

Sears, C.R., Hino, Y., & Lupker, S.J. (1995). Neighborhood size and neighborhood frequency effects in word recognition. *Journal of Experimental Psychology: Human Perception and Performance, 21*, 876–900.

Shallice, T., & Saffran, E. (1986). Lexical processing in the absence of explicit word identification: Evidence from a letter-by-letter reader. *Cognitive Neuropsychology, 4*, 429–458.

Warrington, E.K., & Shallice, T. (1980). Word-form dyslexia. *Brain, 103*, 99–112.

Waters, G.S., & Seidenberg, M.S. (1985). Spelling-sound effects in reading: Time-course and decision criteria. *Memory and Cognition, 13*, 557–572.

Wheeler, D.D. (1970). Processes in word recognition. *Cognitive Psychology, 1*, 59–85.

APPENDIX A

Stimulus List Used in Exp. 1

Target	Target Frequency	Repeated Prime	Unrelated Prime	Target	Target Frequency	Repeated Prime	Unrelated Prime
ACRE	High	AcRe	BaNd	BANG	Low	BaNg	wEed
BAND	High	bAnD	gRaY	BARD	Low	BaRd	dReG
BASE	High	BaSe	tEaR	BARK	Low	bArK	PeAr
BEND	High	bEnD	dAtA	BEAD	Low	BeAd	cAgE
CARD	High	CaRd	hEaR	BEAN	Low	BeAn	rApE
CARE	High	cArE	fEeD	BREW	Low	bReW	jAdE
DARE	High	dArE	BeNd	CAGE	Low	CaGe	bEaD
DARK	High	DaRk	rEaD	DAME	Low	DaMe	bEaN
DATA	High	DaTa	aCrE	DARN	Low	dArN	ReEk
DEAD	High	dEaD	gAvE	DART	Low	dArT	rEeF
DEEP	High	dEeP	HaRd	DEED	Low	dEeD	gOrE
DRAW	High	dRaW	cArD	DEEM	Low	dEeM	GaRb
EVER	High	eVeR	DaRe	DEER	Low	dEeR	wAde
FEED	High	FeEd	MaDe	DREG	Low	DrEg	sAgE
GATE	High	GaTe	DeAd	EDEN	Low	EdEn	bArD
GAVE	High	GaVe	hEaD	GARB	Low	gArB	DeEd
GRAY	High	GrAy	RaCe	GORE	Low	GoRe	DeEm
HARD	High	hArD	DeEp	HEED	Low	HeEd	mArE
HEAD	High	HeAd	gAtE	JADE	Low	JaDe	sEaR
HEAR	High	HeAr	bAsE	MARE	Low	MaRe	ReEd
HERE	High	HeRe	rOaD	PEAR	Low	pEaR	BaRk
MADE	High	mAdE	EvEr	RAKE	Low	rAkE	eDeN
NEED	High	NeEd	dArK	RAPE	Low	RaPe	hEeD
PAGE	High	pAgE	DrAw	REED	Low	rEeD	dAmE
RACE	High	rAcE	nEeD	REEF	Low	ReEf	Dart
RATE	High	RaTe	sEeD	REEK	Low	rEeK	DaRn
READ	High	ReAd	CaRe	SAGE	Low	SaGe	DeEr
ROAD	High	RoAd	hErE	SEAR	Low	SeAr	RaKe
SEED	High	SeEd	rAtE	WADE	Low	WaDe	BrEw
TEAR	High	TeAr	PaGe	WEED	Low	WeEd	bAnG

Note that primes are each shown only under one of the versions used across repetitions of prime–targets pairs, with the other versions corresponding to an inversion of the upper- and lower-case assignments for letters in the prime.

APPENDIX B

Stimulus List Used in Exp. 2

Target	Repeated Prime	Homophone Prime	Unrelated Prime	Target	Repeated Prime	Homophone Prime	Unrelated Prime
ALTAR	altar	alter	beech	MEET	meet	meat	sail
ALTER	alter	altar	creek	PAIL	pail	pale	feat
BAIL	bail	bale	prey	PALE	pale	pail	weak
BALE	bale	bail	seem	PRAY	pray	prey	sell
BEACH	beach	beech	steel	PREY	prey	pray	cell
BEECH	beech	beach	alter	RAIN	rain	rein	tied
BLEW	blew	blue	meat	REIN	rein	rain	bail
BLUE	blue	blew	gait	ROAD	road	rode	heel
CELL	cell	sell	pray	RODE	rode	road	feet
CREAK	creak	creek	alter	SAIL	saiL	sale	flea
CREEK	creek	creak	steal	SALE	sale	sail	blew
FEAT	feat	feet	mail	SEAM	seam	seem	tale
FEET	feet	feat	male	SEEM	seem	seam	gate
FLEA	flea	flee	pail	SELL	sell	cell	tide
FLEE	flee	flea	tail	STEAL	steal	steel	beach
GAIT	gait	gate	blue	STEEL	steel	steal	creak
GATE	gate	gait	heal	TAIL	tail	tale	seam
HEAL	heal	heel	rain	TALE	tale	tail	flee
HEEL	heel	heal	pale	TIDE	tide	tied	meet
MAIL	mail	male	week	TIED	tied	tide	rode
MALE	male	mail	rein	WEAK	weak	week	bale
MEAT	meat	meet	sale	WEEK	week	weak	road

APPENDIX C

Stimulus List Used in Exp. 3

Target	Frequency	Orthographic Neighbourhood Size	Target	Frequency	Orthographic Neighbourhood Size
ACRE	low	Low	SILO	Low	Low
ARCH	low	Low	THUD	Low	Low
BLUR	low	Low	THUG	Low	Low
CHAR	low	Low	TOMB	Low	Low
CHEF	Low	Low	TROT	Low	Low
CYST	Low	Low	VEER	Low	Low
DUKE	Low	Low	VOID	Low	Low
EARL	Low	Low	WATT	Low	Low
EDEN	Low	Low	WEPT	Low	Low
FERN	Low	Low	WIRY	Low	Low
FETE	Low	Low	WISP	Low	Low
FRET	Low	Low	WITS	Low	Low
FROG	Low	Low	BALE	Low	High
FUME	Low	Low	BEAD	Low	High
FUSE	Low	Low	BOOT	Low	High
FUSS	Low	Low	BULL	Low	High
GENE	Low	Low	CAKE	Low	High
GLEN	Low	Low	CAVE	Low	High
GREY	Low	Low	COKE	Low	High
HAWK	Low	Low	CONE	Low	High
JADE	Low	Low	DAME	Low	High
JOWL	Low	Low	DANE	Low	High
KELP	Low	Low	DENT	Low	High
LIED	Low	Low	DINE	Low	High
LIMB	Low	Low	DOLE	Low	High
LISP	Low	Low	DUCK	Low	High
LOAF	Low	Low	FAKE	Low	High
NORM	Low	Low	FOLD	Low	High
OATS	Low	Low	FORE	Low	High
OILY	Low	Low	GALE	Low	High
OXEN	Low	Low	GALL	Low	High
PITY	Low	Low	GORE	Low	High
PONY	Low	Low	HACK	Low	High
PREY	Low	Low	HARE	Low	High
PROD	Low	Low	HEAL	Low	High
ROMP	Low	Low	HOOT	Low	High
ROSY	Low	Low	HOSE	Low	High
SEWN	Low	Low	LACE	Low	High

LAME	Low	High	MUCH	High	Low
LASH	Low	High	NEWS	High	Low
LENT	Low	High	ONCE	High	Low
LICE	Low	High	ONLY	High	Low
LONE	Low	High	OPEN	High	Low
LOOT	Low	High	OVER	High	Low
LORE	Low	High	PLAN	High	Low
LUST	Low	High	PLAY	High	Low
MASH	Low	High	SIZE	High	Low
MOLE	Low	High	SUCH	High	Low
NAIL	Low	High	THEY	High	Low
PATE	Low	High	THIS	High	Low
PEAR	Low	High	THUS	High	Low
RAKE	Low	High	TOWN	High	Low
RAVE	Low	High	TRUE	High	Low
REED	Low	High	TYPE	High	Low
SAGE	Low	High	UNIT	High	Low
SEAR	Low	High	UPON	High	Low
SLOT	Low	High	USED	High	Low
TAME	Low	High	VARY	High	Low
VALE	Low	High	VIEW	High	Low
VEST	Low	High	WALK	High	Low
WALE	Low	High	WAYS	High	Low
WART	Low	High	WHAT	High	Low
ABLE	High	Low	WHEN	High	Low
ALSO	High	Low	WHOM	High	Low
AREA	High	Low	WITH	High	Low
AWAY	High	Low	WONT	High	Low
BLUE	High	Low	BACK	High	High
BODY	High	Low	BALL	High	High
BOTH	High	Low	CARE	High	High
CITY	High	Low	CASE	High	High
CLUB	High	Low	COLD	High	High
DATA	High	Low	COME	High	High
DOES	High	Low	CORE	High	High
DOWN	High	Low	DATE	High	High
EACH	High	Low	DEAL	High	High
ELSE	High	Low	DONE	High	High
EVEN	High	Low	FALL	High	High
FREE	High	Low	FEAR	High	High
FROM	High	Low	FILE	High	High
GIRL	High	Low	FINE	High	High
HIGH	High	Low	FIRE	High	High
INTO	High	Low	FULL	High	High
KEPT	High	Low	GAME	High	High
MANY	High	Low	GAVE	High	High

HARD	High	High	NEAR	High	High
HART	High	High	PAST	High	High
HAVE	High	High	RACE	High	High
HEAD	High	High	RATE	High	High
HOLD	High	High	READ	High	High
HOLE	High	High	ROLE	High	High
LACK	High	High	SALE	High	High
LAST	High	High	SAME	High	High
LATE	High	High	SENT	High	High
LEAD	High	High	TAKE	High	High
LINE	High	High	WALL	High	High
LOST	High	High	WAVE	High	High
LOVE	High	High	WENT	High	High
MAKE	High	High	WIDE	High	High
MALE	High	High	YEAR	High	High
MORE	High	High			
MUST	High	High			

PURE ALEXIA AND THE VIEWING POSITION EFFECT IN PRINTED WORDS

Marie Montant and Tatjana A. Nazir
Centre de Recherche en Neurosciences Cognitives, CNRS, Marseille, France

Michel Poncet
Service de Neurologie et de Neuropsychologie, CHU Timone, Marseille, France

In the present article, we investigated the reading ability of CP, a pure alexic patient, using an experimental paradigm that is known to elicit the viewing position effect in normal readers. The viewing position effect consists of a systematic variation of word recognition performance as a function of fixation location within a word: Word recognition is best when the eyes fixate slightly left from the word centre and decreases when the eyes deviate from this optimal viewing position. A mathematical model (Nazir, O'Regan, & Jacobs, 1991), which provides a good description and quantification of the prototypical shape of the viewing position effect, served to interpret CP's reading performance. The results showed that, like normal readers, CP was able to process all letters of a word in one fixation. However, in contrast to normal readers, reading performance was optimal when CP was fixating the right half of the word. This somewhat abnormal pattern of performance was due to (1) poor perceptual processing in the right visual field, and (2) poor processing of letters situated towards the end of the word, independent of visual field presentation. A similar pattern of performance was obtained with normal readers under experimental conditions in which lexical knowledge was of restricted use. We suggest that CP's reading impairment stems from a dysfunction in the coupling between incoming visual information and stored lexical information. This dysfunction is thought to uncover a prelexical level of word processing, where letter information is weighted differently as a function of letter position in a word-centred space.

Requests for reprints should be addressed to Marie Montant or Tatjana Nazir, Equipe Cerveau et Langage, Centre de Recherche en Neurosciences Cognitives, Centre National de la Recherche Scientifique, 31, chemin Joseph Aiguier, 13402 Marseille cedex 20, France Tel: (33) 04 91 16 44 91; Fax: (33) 04 91 77 49 69; E-mail: montant or nazir@lnf.cnrs-mrs.fr).

This article is dedicated to Jean Requin.

The research reported in this article was supported by a grant from the French Research Department (Ministère de la Recherche et de l'Enseignement supérieur) to M. Montant, a grant from the German Research Office (DFG-Deutsche Forschungsgemeinschaft) and a grant from La Fondation pour la Recherche Medicale to T. Nazir. We are grateful to Art Jacobs, Arnaud Rey, and Johannes Ziegler for valuable advice, support, and helpful comments on an earlier version of this article. We thank Marlene Behrmann, Max Coltheart, and two anonymous reviewers for their constructive suggestions. We also thank Marie-Anne Henaff and François Michel for their evaluation of CP's vision and Frédérique Etcharry for conducting the neuropsychological investigations. We are particularly grateful to CP for his patience.

INTRODUCTION

Pure alexia or letter-by-letter reading is a reading disorder that generally occurs after cortical damage in the left temporo-occipital region. This syndrome is characterised by disproportionately slow reading of single words and text in the absence of other marked neurobehavioural impairments in premorbidly literate adults. The hallmark of pure alexia is the word length effect: reading latencies increase as the number of letters in words increases (e.g. Patterson & Kay, 1982; Reuter-Lorenz & Brunn, 1990; Warrington & Shallice, 1980).

Several hypotheses have been formulated to account for the length effect of pure alexia. Although most investigators agree that pure alexia is related to a loss of the capacity to decode arrays of letters in parallel, disagreement arises with respect to the precise locus of the impairment. Two hypotheses have been proposed. The first states that pure alexia is the result of a visuoperceptual impairment that is not limited to the perception of printed words during reading but is merely most visible in this context (Behrmann & Shallice, 1995; Farah & Wallace, 1991; Hanley & Kay, 1992; Rapp & Caramazza, 1991; Sekuler & Behrmann, 1996). The second states that pure alexia is the result of a selective dysfunction of the mechanisms exclusively involved in reading. In contrast to the first hypothesis, this hypothesis claims that low-level perceptual processes, even when impaired, are not directly responsible for pure alexia (Friedman & Alexander, 1984; Kinsbourne & Warrington, 1962; Levine & Calvanio, 1978; Patterson & Kay, 1982; Warrington & Shallice, 1980). Nevertheless, the impairment is supposed to take place during early stages of the reading process because high-level processes (e.g. semantics, phonology) are not selectively affected in pure alexia (Shallice & Saffran, 1986; but see Arguin, Bub, & Bowers, this issue). Moreover, some pure alexic patients show evidence of lexical activation, as indicated by the presence of a word superiority effect and implicit reading abilities, which suggests that the word representations in the "mental lexicon" are not damaged (Bowers, Bub, & Arguin, 1996; Bub, Black, & Howell, 1989; Coslett & Saffran, 1989; Coslett, Saffran, Greenbaum, & Schwartz, 1993; Levine & Calvanio, 1978; Reuter-Lorenz & Brunn, 1990). Processes involved in reading may therefore be disrupted either at the level of letter representations (Arguin & Bub, 1993; Behrmann, Plaut, & Nelson, this issue; Behrmann & Shallice, 1995; Bub & Arguin, 1995; Howard, 1991; Kay & Hanley, 1991; Levine & Calvanio, 1978; Saffran & Coslett, this issue) or at a level referred to as the "visual word form system", at which orthographic information of an input string is computed prior to semantic analysis (Carr & Posner, 1994; Déjerine, 1892; Posner & Carr, 1992; Warrington & Shallice, 1980). According to Behrmann et al. (this issue), a deficit in letter processing is causal for letter-by-letter reading; however, this deficit does not preclude lexical information to be partially accessed and lexical effects as well as implicit reading abilities emerge from the residual functioning of the interactive system that supported normal reading premorbidly. As an alternative, Saffran and Coslett (this issue) propose to resolve the apparent contraction between impaired letter processing and implicit reading in terms of two reading systems. The first system operates in the damaged left hemisphere and is responsible for explicit laborious identification of words. The second system operates in the right hemisphere and supports fast covert reading.

The Present Study

To describe further the nature of the reading deficit that characterises pure alexia, in the present study we investigated the reading ability of a pure alexic patient within an experimental paradigm that has been shown to elicit an idiosyncratic pattern of reading performance in normal readers. This paradigm consists of measuring recognition performance for briefly presented words while the eyes are fixating different locations in the word (the experimental technique is illustrated in Fig. 1). Under such experimental conditions, a viewing position effect is obtained for normal readers: Word recognition performance is best when the word is fixated slightly left of its centre and decreases as fixation position deviates either leftwards or rightwards from this "optimal viewing position". Figure 2 gives a characteristic viewing position curve obtained in a word identification task for seven-letter words. The viewing position effect is observed for short as well as for long words and generalises over different alphabetic languages and reading tasks (e.g. Brysbaert & d'Ydewalle, 1988; Brysbaert, Vitu, & Schroyens, 1996; Farid & Grainger, 1996; Nazir, 1993; Nazir, Heller, &

Fig. 1. The paradigm of the variable viewing position in words. A fixation point appears at the centre of the computer screen. After a short duration, the fixation point is replaced by a word. A brief exposure duration of the word is adopted to prevent participants from making eye movements. The word appears at different positions relative to the fixation point, such that the directly fixated part of the string can systematically be manipulated from trial to trial. Eye movements are not measured.

Sussman, 1992; Nazir, Jacobs, & O'Regan, in press; Nazir et al., 1991; O'Regan & Jacobs, 1992; O'Regan, Lévy-Schoen, Pynte, & Brugaillère, 1984). A mathematical model, which provides a good description and quantification of the prototypical shape of the viewing position curve (Nazir et al., 1991), served to interpret the deviating reading performance of the patient. The model is described next.

A Model to Account for the Viewing Position Effect

Given the strong acuity drop-off in parafoveal vision, the number of letters that benefit from high resolution differs considerably as a func-

these calculations, unless the letter cluster is very large, orthographic structure does not appreciably change the shape of the viewing position curve. For mathematical tractability, we will work with the independence assumption (see Nazir et al., in press).

For any fixation location and string length, theoretical probabilities of correct word recognition can be estimated by equation:

$$P_{word}(f, a, b_{left}, b_{right}, l) = \begin{cases} \left(\prod_{n=1}^{l-f}(a - n \cdot b_{right})\right) \cdot a & : f = 1 \\ \left(\prod_{n=1}^{f-1}(a - n \cdot b_{left})\right) \cdot a \cdot \left(\prod_{n=1}^{f-1}(a - n \cdot b_{left})\right) \cdot a \\ \left(\prod_{n=1}^{l-f}(a - n \cdot b_{right})\right) & : 1 < f < l \\ & : f = l \end{cases}$$

(1)

where $P_{word}(f, a, b_{left}, b_{right}, l)$ is the probability to recognise a word as a function of f, the relative location of the fixated letter in the string (in letter units); a, the probability to recognise the directly fixated letter; b_{left} and b_{right} the drop-off rate of the probability to recognise a letter with increasing eccentricity, going leftwards and rightwards, and l, the length of the word.

Factors affecting the Viewing Position Effect

According to the model, manipulation of the legibility of letters in words will affect the shape of the viewing position curve in a systematic way. If instead of .05, as in Table 1, parameter b_{right} drops by a value of .1 per letter of eccentricity (which indicates a decrease of the relative legibility of letters), the height of the viewing position curve is predicted to drop, and the viewing position effect (i.e. the shape of the curve) is significantly strengthened. With a drop-off rate of .02 or even 0 (which indicates that independently of their eccentricity, all letters of the word are identified equally well), the curve becomes progressively flat (see Fig. 3a, for these different examples). By manipulating the drop-off rate of letter legibility, either through variations of the inter-letter spaces in words or by increasing letter size proportional to eccentricity, Nazir et al. (1992, in press) showed that these manipulations indeed changed the viewing position curve in a way that was predicted by the model. When letter legibility decreased, the viewing position effect was strengthened (see Fig. 3b, left panel); when letter legibility increased, the effect disappeared (see Fig. 3b, right panel).

Whereas the legibility of individual letters changes the shape of the viewing position curve, variables that affect the word as a unit are additive to the viewing position effect. The frequency of occurrence of a word, for example, changes the total height of the viewing position curve without affecting its shape (e.g. McConkie, Kerr, Reddix, Zola, & Jacobs, 1989;

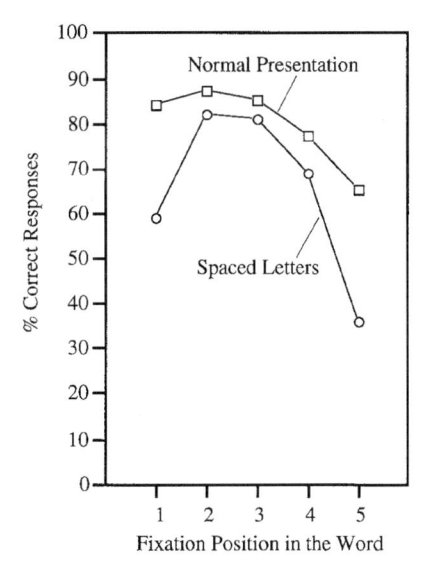

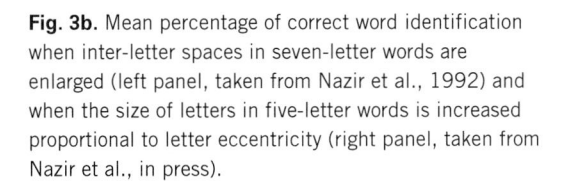

Fig. 3b. Mean percentage of correct word identification when inter-letter spaces in seven-letter words are enlarged (left panel, taken from Nazir et al., 1992) and when the size of letters in five-letter words is increased proportional to letter eccentricity (right panel, taken from Nazir et al., in press).

O'Regan & Jacobs, 1992; Vitu, 1991). Visual and lexical factors seem therefore to affect different parameters of the viewing position curve.

Note that, theoretically, the presence of a viewing position effect implies that all the letters of the string have been processed in one fixation: Variations of recognition performance as a function of the viewing position emerge as the result of the multiplication of the individual letter probabilities in the letter string. This point is important with respect to pure alexia. If pure alexic patients are unable to process all letters of a word rapidly, as normal readers do, they should not show the typical viewing position effect.

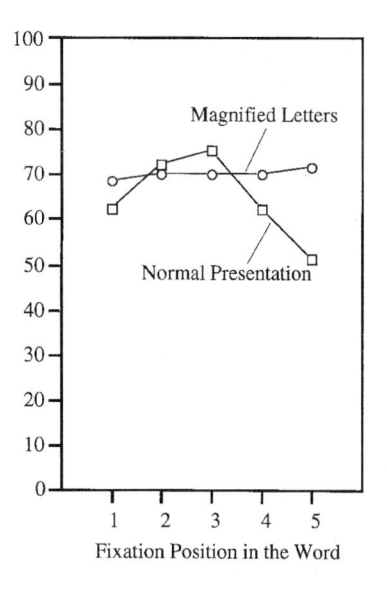

Predictions

A letter-by-letter reading strategy that consists of a left-to-right analysis of the letter string should give better performance with fixations closer to the beginning of words because such fixations would allow immediate letter processing without previous reorientation of attention in search of the first letter. Alternatively, if the system that supports reading is severely damaged, the patient might compensate his reading deficit by referring to local salient features in the words or to letter clusters. If these salient features or letter clusters are randomly distributed in words, no viewing position effect should be expected on performance averaged over a list of different words. By contrast, if the system that supports normal reading is functioning at least partially (as proposed by Behrmann et al., this issue), we should find a viewing position curve that differs from that of normal readers only by shape or height. A difference in shape would point to a perceptual deficit whereas a difference in height would point to a lexical deficit (see Fig. 4 for illustration of these different predictions).

CASE REPORT

CP, a 52-year-old French teacher of physical education at a High School in the south of France, was admitted to the hospital in Marseille after having shown some evidence of a neurological disease on 1 August 1994. During the first hours following his admission, CP demonstrated a right hemiparesis predominantly of the anterior limb and the face and complained of a severe difficulty in finding words. A CT scan and a MRI examination performed 6 days after his admission revealed a hypodense region affecting the left lingual gyrus and the subcortical white matter of the calcarine area, due to an infarct of the left posterior cerebral artery. The splenium of the corpus callosum was spared. The patient exhibited difficulties in retrieving words and a total inability to recognise written language signs ("This is not written as it used to be").

According to a test for handedness (Crovitz & Zener, 1962), CP was ambidextrous. He scored 47 points on a scale where 42 points stands for pure ambidextrality and 70 for left-handedness. Perimetric testing showed that both visual fields were spared: The threshold for detecting a flashed light dot was .48, .53, .57, .8 cd/m^2 for 1, 2, 3 and 5 degrees of eccentricity respectively in the left visual field, and .5, .54, .93, and .8 cd/m^2 for 1, 2, 3 and 5 degrees of eccentricity respectively in the right visual field. Colour-matching tasks revealed that, under brief exposure duration, CP was better at discriminating colour discs presented in the left visual field (32/32 correct) than in the right visual field (26/32 correct). Neurolinguistic investigations performed during the week following admission demonstrated that language comprehension, writing, verbal fluency, and sentence repetition were entirely spared. However, performance was affected in letter naming (44/50 correct) and visual lexical decision tasks (47/72 correct). Errors in the lexical decision task were essentially due to a tendency to reject words, independent of their frequency of occurrence. Reading aloud was laborious, letter-by-letter or syllable-by-syllable. Most errors were visual in nature (e.g. the French word "fete" was pronounced /jet/), some were regularisations (e.g. the French word "gars" was pronounced /gars/ instead of /ga/). Reading words that were spelled aloud to him occasionally resulted in regularisation errors (e.g. the French words "s,e,p,t" was pronounced /sɛpt/ instead of /sɛt/). CP made no errors in writing irregular words to dictation.

One month later, the motor function of the right anterior limb and hemiface had fully recovered. The repetition of the same neuropsychological testing also revealed that, after one month, CP's letter naming performance was comparable to that of normals (no errors) and his lexical decision performance had considerably improved (69/72). In the lexical decision task, CP had apparently adopted the strategy of writing the stimuli before responding. Reading aloud was still very slow, and

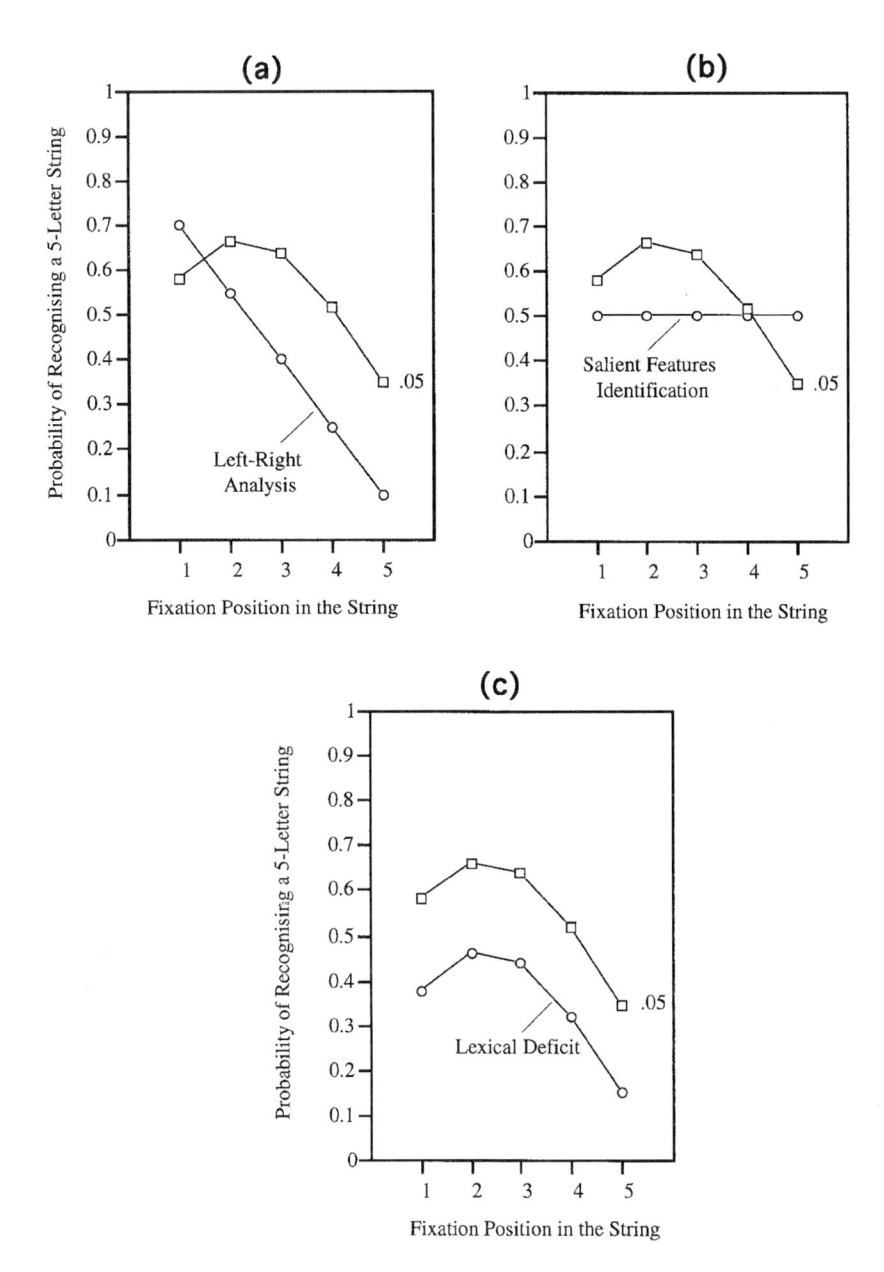

Fig. 4. Theoretical probabilities of recognising a five-letter string as a function of fixation position in the string predicted by the model for a pure alexic patient. The viewing position curve (a) should be linearly decreasing towards the right if the patient uses a left-to-right analysis of the word; (b) it should remain flat if the patient recognises words using salient features; and (c) the height of the curve should decrease if the patient has a lexical deficit.

visual and regularisation errors were occasionally observed. Before starting our investigations, we asked the patient to read aloud 20 words among which 10 were irregular words. CP made no regularisation errors. We tested CP over the course of 1 year, starting November 1994. The patient gave his informed consent to participation in these experiments.

EXPERIMENT 1

To assess CP's reading ability, we used a word identification task in combination with the technique of the variable viewing position. The aim of Experiment 1 was to determine whether CP showed a typical viewing position effect.

Method

Materials

The experimental stimuli were 250 five- to nine-letter French words (50 stimuli per length). The words were selected from *Brulex*, a computerised lexical database that provides orthographic, phonological, and grammatical information for a corpus of 36,000 French words along with their frequency of occurrence and other lexical statistics (Content, Mousty, & Radeau, 1990). Word frequency ranged from 10 to 1476 occurrences per million with an average frequency of 41 occurrences per million. Words of different lengths were matched as closely as possible for frequency. In order to reduce the possibility of visual/orthographic errors to occur, we only selected "hermits", i.e. words without orthographic neighbours[2]. No verbs were used. All stimuli are listed in Appendix A.

Apparatus

Stimuli were presented in the centre of a monochrome screen of an Apple Macintosh Powerbook 145 microcomputer, refreshed at 60Hz. The words were displayed in capital letters, black fonts against a white background, in the standard Macintosh font Courier (size 24 points). Each letter was defined in a pixel matrix of 4 × 5mm. At a viewing distance of about 60cm, a five-letter word subtended 2.87 degrees of visual angle.

Design

Each word was divided into five equally wide zones of one fifth of the word length (e.g. 1 letter wide for five-letter words, 1.8 letters wide for nine-letter words). The centre of each zone was designated as a potential initial fixation position. This was done to obtain comparable fixation positions in all words, independent of their length. The set of 50 words of a given length was divided into 5 groups of 10 words each. Each word group was presented at one of the 5 different fixation

---

[2] Orthographic neighbours are words of the same length that differ in one letter (for example, "train" and "brain"). It has been shown that orthographic neighbourhood manipulations affect the performance of normal participants during word recognition (for a review, see Grainger & Jacobs, 1996).

positions. Testing was repeated over 5 experimental sessions with the same 250 words. The attribution of a particular fixation position to a particular word group was done differently for each experimental session, such that, taken over all sessions, each word was seen from all 5 fixation positions. Stimuli were presented in blocks of the same word length, the order of the blocks was counterbalanced across experimental sessions following a Latin square design. The 50 stimuli in a block were presented in random order. There was an interval of at least 3 weeks between 2 experimental sessions.

Procedure

A fixation point (:) appeared at the centre of the screen for 1 second. It was replaced by a word that remained on the screen for 183msec (exposure duration was adjusted during repeated sessions of 50 practice trials until the percentage of correct responses exceeded 30%; practice words were different from test words). The word was laterally shifted with respect to the fixation point such that, on its appearance, the middle of one of its five fixation zones was fixated in (see Fig. 1). The word was replaced by a backward mask that consisted of a string of hash marks. The mask exceeded the stimulus by one character on each side. The patient was asked to pronounce the word, or, if that was not possible, to spell aloud the letters he had seen. The mask remained on the screen until the patient pressed a key that triggered the next trial. The experimenter registered the responses. No feedback was given. Each session of the experiment was preceded by 10 practice trials. Practice words were different from test words. Naming latencies were not measured: Percentage of correct word identification was the dependent variable.

Fifteen undergraduate students from the University of Marseille participated in a control experiment. All were native French speakers and had normal or corrected-to-normal vision. Participants were paid for their participation. Experimental conditions were identical to that of the patient except for the following: The same 250 words as used for CP and additional 250 words of comparable characteristics served as stimuli (word frequency ranged from 10 to 1476 occurrences per million with an average frequency of 49 occurrences per million). Presentation duration was 33msec. The experiment was run on a microcomputer Macintosh IIci.

Results and Discussion

To start with, we illustrate that CP showed the classic word length effect as is indicative for pure alexia. Figure 5 provides the mean percentage of correct word identifications for CP and normal participants as a function of word length. Note that, in this figure, data are not collapsed across the five fixation positions but correspond to the condition in which words were fixated at the centre. A full logistic linear regression showed that CP's performance dropped by 12% per additional letter ($y = -12x + 134$; $r^2 = .45$; $P < .001$). In contrast, performance of normal participants dropped by 2.79% per additional letter ($y = -2.97x + 85.83$; $r^2 = .04$, $P = .07$). Thus, the length effect was

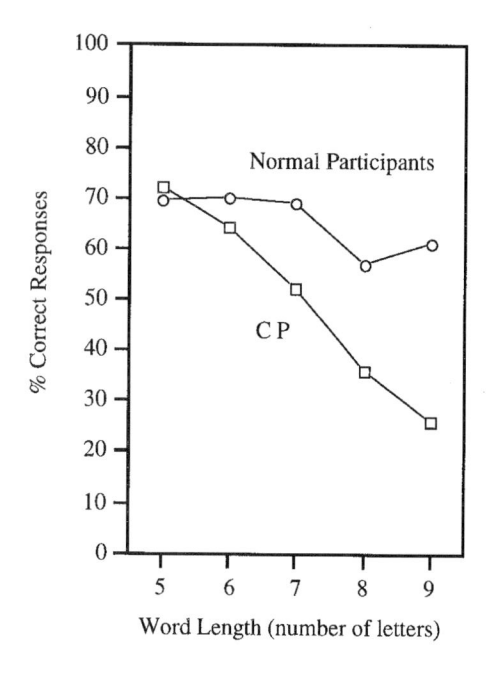

Fig. 5. Mean percentage of correct word identification as a function of word length for CP and normal participants in Experiment 1. Data correspond to the condition in which words were fixated at the centre.

approximately four times bigger for CP than for normal participants, if we consider that normal participants showed a length effect at all (note that it was not systematic).

The left panel of Fig. 6 gives CP's mean percentage of correct word identification for the different word lengths as a function of the five fixation positions in words. The right panel provides comparable results for the controls. An analysis of variance (ANOVA) with session as random variable showed that, for all word lengths, CP had a clear viewing position effect [$F(4,16) = 21.47, P < .001; F(4,16) = 10.61, P < .001; F(4,16) = 9.21, P < .001; F(4,16) = 3.91, P < .05; F(4,16) = 2.91, P = .055$ for 5- to 9-letter words respectively]. Normal participants also showed a viewing position effect that was significant in an ANOVA with subjects as random variable [$F(4,56) = 28.66, P < .001; F(4,56) = 39.12, P < .001; F(4,56) = 64.64, P < .001; F(4,56) = 71.03, P < .001; F(4,56) = 51.20, P < .001$ for five- to nine-letter words, respectively]. However, the shape of CP's viewing position curve differed from the one typically observed for normal participants. Whereas performance of normal participants was best when words were fixated slightly left of their centre, performance of the patient was best when words were fixated towards their end. None of the controls showed a maximum of performance while fixating the right part of words. A simple regression analysis showed that CP's overall performance improved across sessions ($y = 4.3x + 19.7; r^2 = .88; P < .05$) but the shape of the viewing position effect was nearly constant across sessions. At every fixation position, CP exhibited a clear length effect, except when performance was extremely low. Note, however, that performance of the control group was sensitive to word length as well, in particular when words were fixated towards their ends ($y = -7.3x + 92.93, r^2 = .18, P < .001; y = -6.9x + 65.17, r^2 = .26, P < .001$ for fixation positions 4 and 5 respectively). An unpaired post-hoc t-test showed that CP's performance, like that of normal participants, was highly sensitive to word frequency [$t(248) = 5.27, P < .001; t(434) = 7.37, P < .001$ for CP and controls, respectively]. The size of this frequency effect was 21.1% for the patient and 15.63% for normal participants.

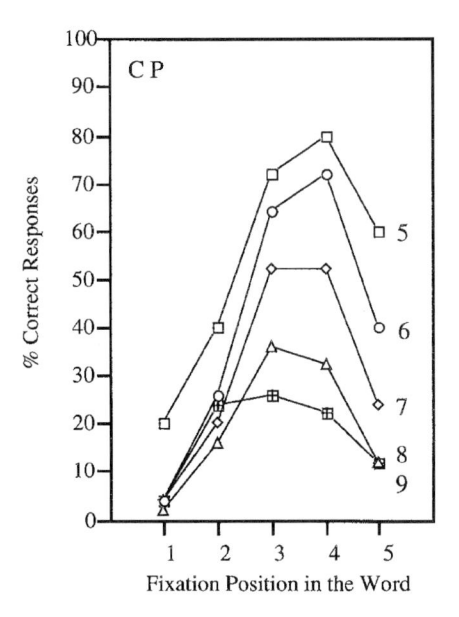

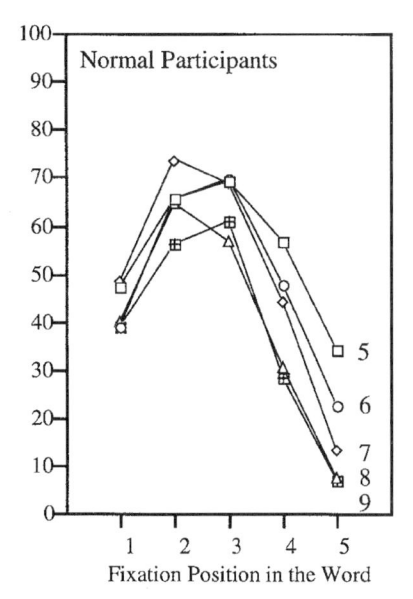

Fig. 6. Mean percentage of correct word identification for CP (left panel) and normal participants (right panel) as a function of word length (five- to nine-letter words) and fixation position in words (Experiment 1). For CP, each data point corresponds to 50 measures.

The present data clearly show that the reading system of the patient is functioning at least partially. CP exhibited a strong frequency effect and a clear viewing position effect, although the shape of the viewing position curve was not of the classic type. Like normal readers, CP was able to process a string of letters during one single fixation, provided that he fixated towards the second half of the word. According to the model, asymmetries in the viewing position curve are caused by differences in the ability to identify letters in the right and left visual field (Nazir et al., 1991). A shift of the optimal viewing position to the right of the centre of the word, as observed with CP, indicates that letter processing is impaired in the right visual field. Given that perimetric testing did not reveal major visual anomalies in CP's right visual field, this impairment cannot stem from a pure visual deficit but must be related to difficulties in processing complex visual forms.

EXPERIMENT 2

To discern potential differences in CP's general capacity to process complex visual stimuli in the right visual field and the left visual field, CP was asked to match the identity of two simultaneously presented letters. The letters consisted of either physically identical pairs (BB), nominally identical pairs (Bb), or nonidentical pairs (BJ). The stimuli were presented either in central vision or in the left or right visual field.

Method

Materials and Design

Thirty-six pairs of letters served as stimuli. These pairs were divided into six groups, corresponding to the six possible combinations of letter identity and letter case: same name/same case for lower and upper case (e.g. bb and BB respectively); same name/different case (e.g. Bb); different name/same case for lower and upper case (e.g. bj and BJ respectively); different name/different case (e.g. Bj). In the same-name conditions, the six letters B, D, G, H, Q, and T served as stimuli. These letters were chosen because of the morphologic dissimilarity of their upper-case and lower-case forms. In the different-name conditions, letters were paired in the following way: B-J, D-F, G-L, H-R, Q-N, and T-M. All pairs are listed in Appendix B. Each pair was presented three times, once at fixation, once to the left, and once to the right of fixation. Thus, one experimental session consisted of 6 × 6 × 3 = 108 trials. Trials were randomised before being equally distributed over two blocks. Testing was repeated over five experimental sessions, with an interval of at least 3 weeks between two sessions. The order of the blocks was counterbalanced across sessions and trials within each block were randomised before each session. Exposure duration varied between sessions; it was either 33msec, 50msec, 67msec, 83msec, or 100msec. The order of these five durations conditions across sessions was chosen randomly. Each session was preceded by six practice trials.

Procedure

Each trial began with the presentation of a fixation point at the centre of the screen. After 1sec, the fixation point was replaced by a pair of letters that remained on the screen for 33, 50, 67, 83, or 100msec. The letters were presented simultaneously, one above the other, and displayed either centrally or 2.87 degrees to the right or left of fixation (this eccentricity corresponds to the width of a five-letter word in Experiment 1). No backward mask was used. The patient was asked to verbally indicate whether the two letters were nominally identical, independent of their physical shape. The experimenter registered CP's responses. No feedback was given. The patient triggered the next trial by pressing a key on the computer keyboard. Reaction times (RTs) were not recorded. The dependent variable was the percentage of correct responses.

Results and Discussion

Figure 7 provides the percentage of CP's correct responses as a function of presentation location and exposure duration. Chi-square tests showed that CP's performance was not different for physical and nominal matches [all $\chi^2(1) < 2.5$], therefore data were collapsed across case conditions in the following analyses (note that differences in physical/nominal matches are usually observed on RTs in pure alexic patients). Performance was better when pairs of letters appeared at fixation than in the left visual field (LVF) [$\chi^2(1) = 5.14, P < .05$] or in the right visual field (RVF) [$\chi^2(1) = 38.19, P < .001$]. In contrast to normal subjects (i.e.

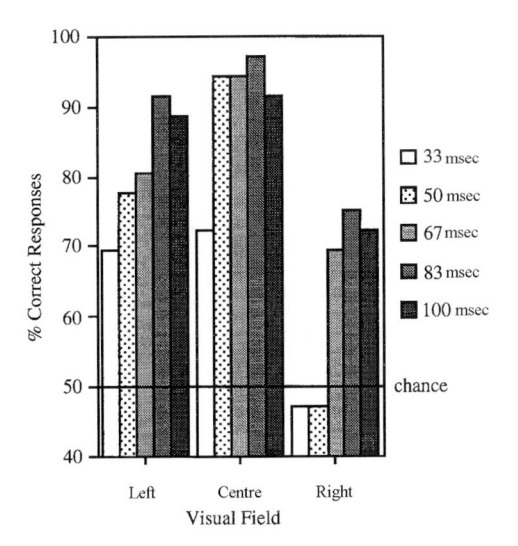

Fig. 7. Mean percentage of correct responses of CP as a function of exposure duration and visual field presentation in Experiment 2. Each bar corresponds to 36 measures.

Bouma, 1971), CP's performance differed between visual fields [$\chi^2(1) = 16.85, P < .001$ for data averaged across all exposure durations]. For exposure duration of 50msec, performance was at chance when letters appeared in the RVF, whereas it was above chance in the LVF [$\chi^2(1) = 4.88, P < .05$]; performance between LVF and RVF was significantly different [$\chi^2(1) = 5.93, P < .05$]. For longer exposure durations, there was a trend towards a difference between visual fields but this difference was not significant [$\chi^2(1) = 0.67, P > .1; \chi^2(1) = 2.5, P > .1; \chi^2(1) = 2.22, P > .1$, for 67, 83, and 100msec of exposure duration, respectively]. Thus, although the patient had no hemianopsia in the strict sense, these results indicate the presence of a deficit in processing relatively complex visual stimuli presented in the RVF.

A general visual disorder, as indicated by these results, will certainly affect reading performance. However, unless its impact on reading can be fully determined, there is no guarantee that this disorder is exclusively responsible for CP's abnormal reading performance.

EXPERIMENT 3

To investigate the impact of CP's right visual field deficit on reading, we tested CP's ability to recognise individual letters within words while words were fixated at the first, middle, or last letter. Letter report accuracy was determined with a two-alternative forced-choice letter identification task (Reicher, 1969). In this task, participants were asked to choose between two letter alternatives, one of which was presented in the word. The target letter could occupy any position in the word. Fixation location and target letter position in words varied independently. Note that when words were fixated at the first letter, all but the first letter appeared in the right visual field. When words were fixated at the last letter, all but the last letter appeared in the left visual field. Thus, CP's ability to process individual letters within words was tested in both visual fields. If a general visual disorder is sufficient to account for CP's abnormal viewing position effect, we should find a target letter position effect in words when letters appear in the right visual field but not when they appear in the left visual field.

In Reicher's original procedure, both response alternatives typically complete a word. Thus, given the stimulus WIND, for example, participants would be asked to determine which of the two alternative target letters N and L was presented at the third letter position in the stimulus. Since both alternatives complete a word (WIND and WILD), participants could not improve their performance by merely guessing a letter that formed a word. In the present experiment, however, only one of the two response alternatives completed a word; the other one led to a pronounceable, orthographically legal pseudoword. Since only words were presented, only letters that completed words were correct responses. This was done to encourage the patient to use lexical knowledge.

Method

Materials and Design
A total of 150 French five-letter words with an average frequency of 131.3 per million (Brulex, Content et al., 1990) served as stimuli. Word stimuli were selected such that the same target letter could be tested across all five letter positions in words. Thus, the target letter D, for example, was tested in the words *d*anse, i*d*ole, or*d*re, gar*d*e, and blon*d*. The letter alternative that led to an incorrect response was not present at any other location in the word. Word frequencies were closely matched across all conditions of target letter position. Among the 150 words used in the experiment, 50 were fixated at the first letter, 50 at the third letter, and 50 at the fifth letter. Among the 50 words of a given fixation condition, 10 words had a target letter at the first letter position, 10 at the second letter position, 10 at the third letter position, 10 at the fourth letter position, and 10 at the fifth letter position. All stimuli are listed in Appendix C. Trials were randomised before being equally distributed over two blocks. Testing was repeated over five identical experimental sessions, with an interval of 2 or 3 weeks between two sessions. The order of the blocks was counterbalanced across sessions and trials within each block were randomised before each session. Each session was preceded by 21 practice trials using words that were different from test words.

Apparatus
The apparatus was the same as in Experiment 1.

Procedure
At the beginning of each trial, a fixation point (:) appeared at the centre of the screen. After 1sec, the fixation point was replaced by a word that remained on the screen for 83msec. The word was displayed in lower-case letters. It appeared at different positions relative to the fixation point, such that the directly fixated letter in the string could be either the first, the third, or the last letter. The word was replaced by a backward mask (a string of seven hash marks), and two response alternatives. The response alternatives were capital letters presented above and below the position of the masked target letter. They were randomly assigned to the above and below positions on each trial. CP was asked to choose the target

letter among the alternatives. Responses were given verbally and noted by the experimenter. The mask and the alternatives remained visible until CP triggered the next trial by pressing the return key. The patient was explicitly informed that all stimuli were words, and no nonword could be presented. RTs were not measured. The dependent variable was the percentage of correct responses.

A control experiment was run with 20 French students (aged between 22 and 30 years). Participants had normal or corrected to normal vision. They were paid for their participation. Materials, design, and procedure were the same as in the experiment described earlier, except for the following modifications: Presentation duration was decreased to 33msec to ensure that performance was not at ceiling, each participant performed only one experimental session, and the experiment was run on a microcomputer Macintosh IIci.

Results and Discussion

In Fig. 8, per cent correct responses of CP and normal participants are presented for the three fixation conditions as a function of the position of the target letter in the word. Data are presented with a 95% confidence intervals around each point (Loftus & Masson, 1994; the computation of CP's confidence intervals was based on item data for repeated sessions).

For normal participants (right panel of Fig. 8), performance varied as a function of fixation position; it was best when words were fixated at the centre, intermediate when fixated at the beginning, and worst when fixated at the end.

This viewing position effect was significant $[F(2,38) = 21.11, P < .001]$. However, within each fixation condition, performance did not differ significantly as a function of target letter position $[F(4,76) = 1.33, P > .1]$.

In comparison to these results, CP showed significant deviations (left panel of Fig. 8). Performance of the patient was maximal for central fixations but, contrary to normal participants, CP performed better when he was fixating the end than the beginning of words $[F(2,8) = 12.94, P < .01]$. This result confirms the existence of a right visual field deficit. However, on top of this, CP's results revealed difficulties in processing letters that were placed in the second half of the word (i.e. letter positions three, four, and five). CP's performance varied significantly as a function of target position $[F(4,16) = 6.8, P < .01]$. This difficulty in reporting final letters remained even when the word was displayed in his intact left visual field [performance for target letters three and four are significantly different from performance for target letters one and two even though all these letters were displayed in the intact left visual field, $F(1,4) = 4.87, P < .05$]. Therefore these difficulties cannot be attributed to the visual disorder identified in Experiment 2. Note that target letter position and fixation position interacted significantly $[F(8, 32) = 3.47, P < .01]$. This was due to the fact that CP's difficulties in processing letters placed in the second half of words could partly be compensated for when visual resolution was high: Performance was higher when the target letter was directly fixated, independent of its position in the string. In short, in addition to a

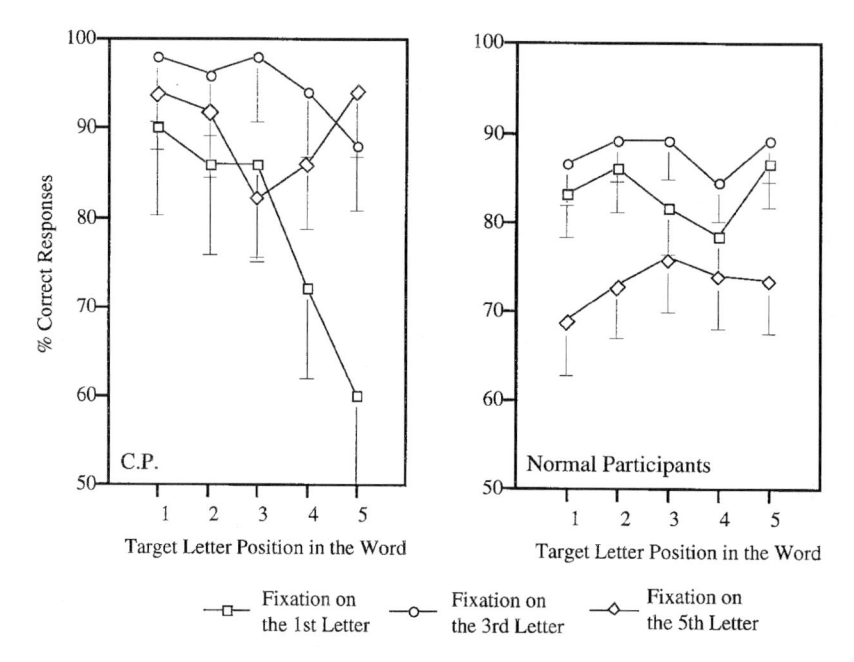

Fig. 8. Mean percentage of correct responses of CP (left panel) and normal participants (right panel) as a function of fixation position and target letter position in words, together with a 95% confidence interval around each data point (Experiment 3). For CP, each data point corresponds to 50 measures.

general visual disorder involving the right visual field, the results of the present experiment suggest that the reading deficit of the patient could also stem from an impairment that consists of a difficulty in processing the right part of words.

Before overinterpreting the present findings, it might be interesting to note that, under certain experimental conditions, normal participants show a pattern of performance similar to the one observed with CP. When lexical knowledge is of limited help in determining target letter identity, that is, when both letter alternatives complete a word of comparable frequency, forced-choice performance is less accurate for target letters that are placed towards the end of words than for letters placed in the first half of words (Montant & Nazir, 1995; for similar results, see Rumelhart & McClelland, 1982; Papp & Johansen, 1994). The present experiment was explicitly designed to encourage participants to use lexical information. According to the dual read-out model of the Reicher paradigm presented by Grainger and Jacobs (1994), correct forced-choice letter report may occur when either the correct letter representation or the correct word representation containing the correct letter reach some critical level of activation. In the context of this dual read-out mechanism, lexical knowledge could be used in Experiment 3 to "infer" the identity of the target letter independent of its position in the word because only one response alternative completed a word and participants knew that nonwords were not presented. As indicated by the absence of a target position effect, normal participants apparently benefited from this manipulation. CP, however, failed to exploit this additional information for the purpose of letter identification. As will be demonstrated in Experiment 4, normal participants and CP behave essentially alike when lexical knowledge is of restricted use.

EXPERIMENT 4

In Experiment 4, we used a lexical decision task and a limited number of repeatedly presented five-letter words and pseudowords. We chose the lexical decision task instead of a classic Reicher task to engage CP in processing the whole letter string. The Reicher task can be performed using single-letter information (even if reading the word was helpful in Experiment 3 because target letter position was unpredictable) whereas to decide whether a letter string is a word or not, the entire string has to be taken into account. The pseudowords on the "no" trials differed from the test words by a single critical letter, which was either the first, middle, or the last letter. This made words and pseudowords very similar. Therefore, unlike in the previous experiment, word read-out alone should not improve performance since participants had to verify the spelling of each stimulus carefully in order to give a correct response. This manipulation was chosen to avoid correct responses being retrieved from lexical knowledge after only a few letters of the string have been identified.

Method

Materials and Design

In the word condition, the five French words AUCUN, AUSSI, ORDRE, ROMAN, and SUJET served as stimuli. Their mean frequency was 496 per million (Brulex, Content et al., 1990); none of these words had orthographic neighbours. For the pseudoword condition, three different pseudowords were constructed from each test word by replacing either the first, the middle, or the last letter of the word with a letter that occurs at the same position in one of the other test words. For example, the pseudowords *R*UCUN, AU*D*UN, and AUCU*T* were constructed from the test word AUCUN. The R in RUCUN occurs at the first position in the test word ROMAN; the D in AUDUN occurs at the middle position in the test word ORDRE, and the T in AUCUT occurs at the last position in the test word SUJET. This was done to avoid that the presence of a certain critical letter in the stimulus could be sufficient to discriminate pseudowords from words. All but one pseudoword met this criterion (stimuli are listed in Appendix D). Pseudowords were orthographically legal, and none of them was a pseudohomophone[3]. Stimuli were displayed

---

[3] The need to control for the homophony of the pseudowords was taken into consideration since a number of studies demonstrated that the phonological form of a letter string affects performance in letter and word identification tasks (e.g. Coltheart, Davelaar, Jonasson, & Besner, 1977; Ziegler & Jacobs, 1995; Ziegler, Van Orden & Jacobs, 1997).

in mixed upper- and lower-case letters to prevent subjects from using global shapes to discriminate between words and pseudowords.

A total of 15 pseudowords were constructed (3 pseudowords from each of the 5 test words). To equalise the number of "yes" and "no" trials, the same five test words were repeated three times. There were three conditions of fixation position: on the first, third, and fifth letter of the string. The 30 items (15 words and 15 pseudowords) were repeated at all 3 fixation positions, resulting in a total of 90 trials. Testing was repeated over four identical experimental sessions, with an interval of 2 or 3 weeks between two sessions. Trials were randomised before each session. A short break occurred halfway through the experiment. Ten practice trials preceded each session.

For personal reasons, the patient only participated in two experimental sessions with these stimuli. He had previously performed two sessions of a pilot experiment that was identical to Experiment 4, except that the identity of the critical letter in pseudowords was not as well controlled as in the main experiment (see Appendix D). Prior analyses showed that the stimuli of these two experiments yielded the same pattern of results. Therefore, data will be averaged across the four sessions in the Results section.

Procedure

Viewing conditions and screen display were the same as in Experiment 3, with the exception that presentation duration was increased to 100msec because CP's performance during the practice trials was too low with shorter exposure durations. The patient was asked to indicate whether the stimulus was a word or not by pressing one of two pre-designated keys on the computer keyboard. The next trial was triggered automatically 1sec after the response. Responses were not corrected. The dependent measure was the percentage of correct responses. RTs were not recorded. The patient knew that words and pseudowords were very similar and extensively repeated; he was explicitly told that identifying all the letters of the stimulus was necessary to discriminate between words and pseudowords.

A control experiment was run with 20 French students (20 to 31 years old). Participants had normal or corrected-to-normal vision. They were paid for their participation. Materials, design, and procedure were the same as in the experiment described above, with the following exceptions: Exposure duration was decreased to 33msec, the experiment was run on a microcomputer Macintosh IIci, and each participant performed only one experimental session.

Results and Discussion

Table 2 gives the percentage of correct responses for CP and normal participants for words and pseudowords in the three fixation conditions as a function of critical letter position. The manipulation of the critical letter position only concerns pseudowords. On the basis of hits and false alarm rates, nonparametric estimates of discriminability (A') were computed for each fixation × critical letter position condition (Grier, 1971). An A' value of 1 indi-

Table 2. Percentage of Correct Responses of CP and Normal Participants for Words and Pseudowords in the Three Fixation Conditions as a Function of Critical Letter Positions

	Fixation Position 1	Fixation Position 3	Fixation Position 5
CP			
Pseudowords			
1st letter critical	90	85	90
3rd letter critical	45	55	50
5th letter critical	30	45	75
Words	58.34	88.34	86.66
Normal Participants			
Pseudowords			
1st letter critical	86	78	81
3rd letter critical	64	47	72
5th letter critical	57	64	72
Words	65.66	76.32	65.00

The manipulation of the critical letter position only concerns pseudowords.

cates perfect discrimination, whereas a value of .5 is equal to random discrimination. A' values for below-chance performance were calculated with Aaronson and Watts' (1987) extension of the A' formula. The A' scores of CP and normal participants are given in Fig. 9.

As can be seen in Fig. 9a, the patient's performance was very similar to that in Experiment 3. As indicated by the difference in the height of the curves, CP showed a reversed viewing position effect: He performed better, independent of critical letter position, when fixating the last letter of a string than the first letter. An ANOVA carried out on A' scores revealed that this effect was significant [$F(2,6) = 31.18, P < .001$]. This viewing position effect confirms the existence of a right visual field deficit. More important, however, independent of where the patient was fixating, performance was better when the critical letter in the string was occupying the first letter position rather than the middle or last letter position [$F(2,6) = 44.24, P < .001$]. Even when the critical letters were presented in the intact left visual field, that is, when fixation was on the fifth letter of the string, performance was better for letters at the first than at the third position [$F(1,3) = 16.35, P < .01$]. As in Experiment 3, critical letter position and fixation position interacted significantly [$F(4,12) = 10.01, P < .001$] because performance for letters at the fifth position increased when these letters were directly fixated.

In contrast to CP, normal participants showed no difference in the overall height of the curves (see Fig. 9b), that is, there was no viewing position effect [$F(2,38) = .5, P > .5$]. However, their performance varied as a func-

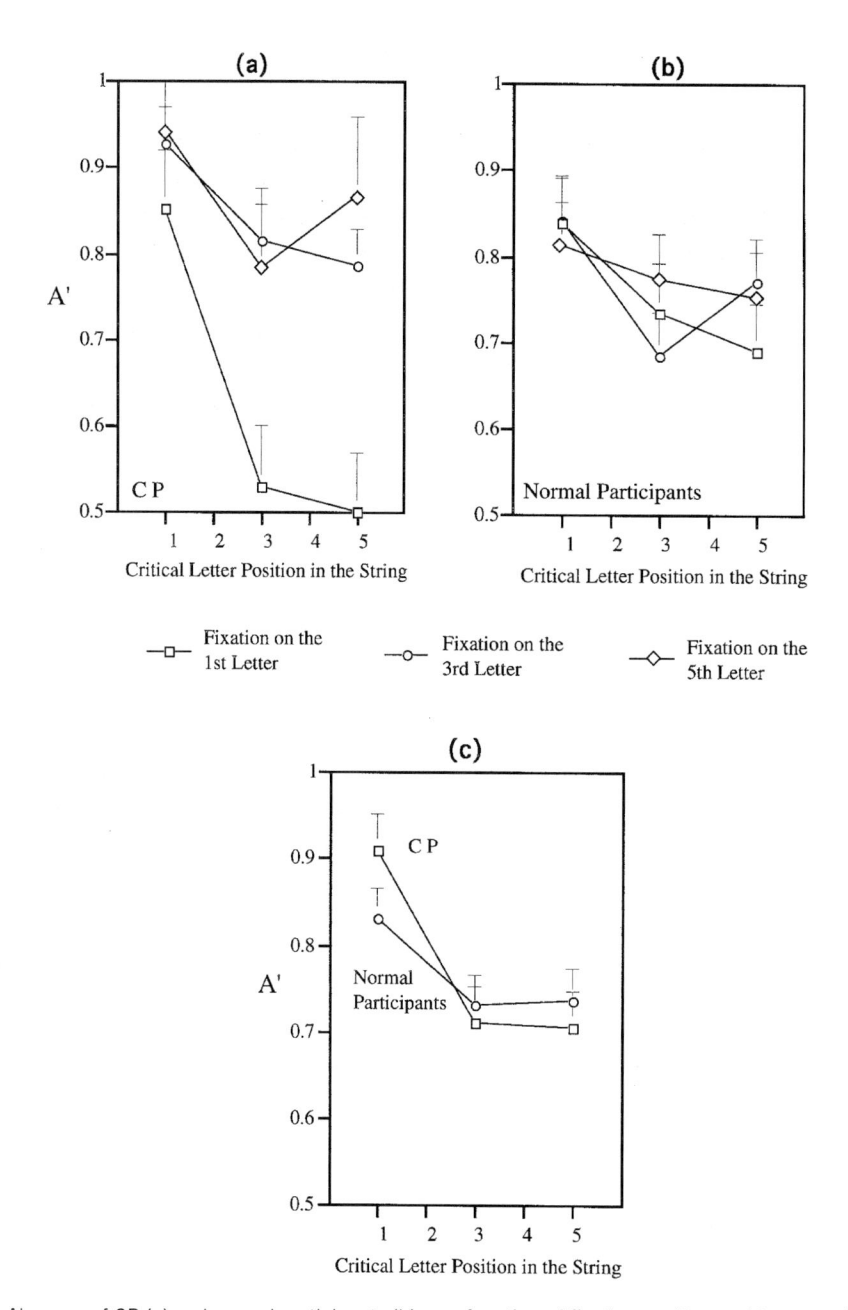

Fig. 9. A' scores of CP (a) and normal participants (b) as a function of fixation position and letter position in the string, together with a 95% confidence interval around each data point (Experiment 4). For CP, each data point corresponds to 120 measures. (c) A' scores, averaged across fixation conditions, for CP and normal participants as a function of critical letter position, together with a 95% confidence interval around each data point.

tion of the critical letter position in pseudowords in a similar way as it did for CP [$F(2,38) = 9.99$, $P < .001$]. Performance was highest for the first letter and lowest for the last letter, independent of where the eyes were fixating. For a better comparison, Figure 9c gives the mean A' values of CP and normal participants averaged over all fixation locations. Thus, normal participants and CP seemed to behave essentially the same in Experiment 4: Lexical decision performance was poor when the critical letter was placed in the second part of the string.

Because one could argue that the experimental paradigm used in Experiment 4 did not capture perceptual mechanisms that form part of normal word identification, we propose to perform additional analyses on the word recognition data of Experiment 1.

Further analyses

To support further the hypothesis that CP has difficulties in processing letters placed in the second half of words, the data of Experiment 1 were re-analysed at the letter level. For this, participants' responses (correct and incorrect) were scored depending on whether a reported letter was present in the displayed word. A response was considered correct independently of report order. We opted for this solution because correct report order is difficult to maintain under rapid visual presentation. Thus, for example, when the patient replied "plante" to the presentation of the target word "PLANETE", the first four letters as well as the final letters T and E were considered correct (even though the position of the final letters did not match the absolute position in the target word). Similarly, when the patient replied "fourmi" to the presentation of the word stimulus "ETOURDI", the letters O, U, R and I were counted as correct responses. The same held for a response like "froumi". This analysis was performed on the data of Experiment 1 for both CP and normal participants.

Figure 10 gives letter identification scores as a function of serial letter position. Data for the different word lengths are presented separately in the different panels. Each panel plots letter recognition scores for the five viewing position conditions. Data are plotted aligned at fixation location, which is indicated by the central vertical line. Thus, the right half in each panel gives scores for letters that fell into the right visual field; the left half gives scores for letters that fell into the left visual field. Performance for CP and for normal participants are plotted respectively in the left and right panels of Fig. 10.

Note that all following analyses were performed for conditions in which the eyes were fixating the beginning or the end of the words only. An ANOVA performed on letter identification scores showed that, for normal participants, performance was consistently better in the RVF than in the LVF for all word lengths [$F(1,14) = 14.68$, $P < .01$; $F(1,14) = 8.18$, $P < .05$; $F(1,14) = 43.23$, $P < .001$; $F(1,14) = 49.39$, $P < .001$; $F(1,14) = 73.14$, $P < .001$ for five-, six-, seven-, eight-, and nine-letter words, respectively). This confirms earlier findings reported in the literature (e.g. Bouma, 1973; Bouma & Legein, 1977; Hagenzieker et al., 1990; Nazir et al., 1991). In both visual fields, letter report

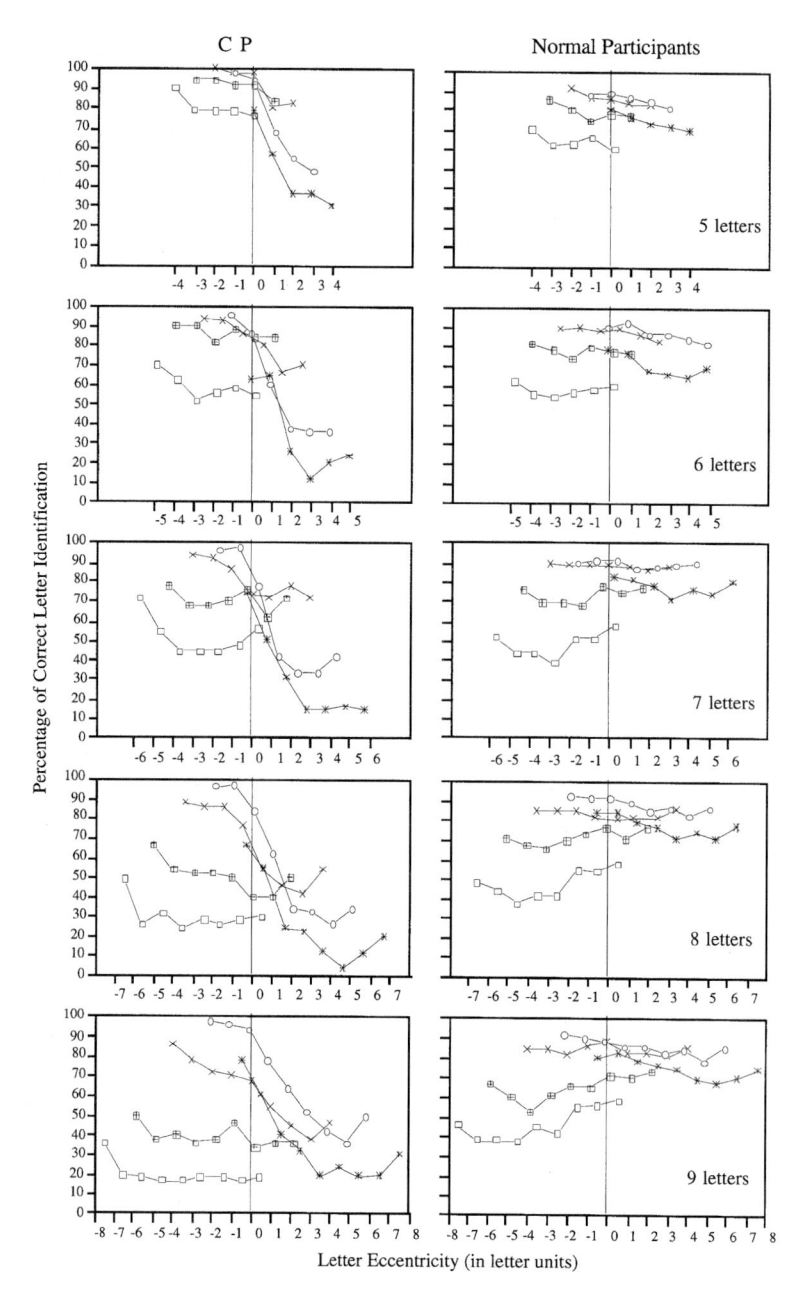

Fig. 10. Percentage of correct letter identification for CP (left panels) and normal participants (right panels) as a function of serial letter position (from Experiment 1). Data for the different word lengths are presented separately in the different panels. Each panel plots letter identification scores aligned at fixation (fixation is indicated by the central vertical line) for the five viewing conditions.

accuracy dropped with increasing distance from fixation location, with a stronger drop-off rate in the LVF. This is most obvious when we compare performance for conditions in which the eyes were fixating the beginning vs. the end of words [the interaction between visual field presentation and serial letter position is significant for five-, six-, and eight-letter words; $F(4,56) = 6.04$, $P < .001$; $F(5,70) = 2.59$, $P < .05$; $F(7,98) = 2.15$, $P < .05$, respectively]. Additionally, the number of letters to be recognised affected performance more dramatically when letters fell into the LVF than into the RVF. A significant word length effect is observed in the LVF [$F(4,56) = 8.54$, $P < .001$] but not in the RVF [$F(4,56) = 1.63$, $P > .1$]. For a given word length, each additional letter presented in the LVF systematically lowered report accuracy of all letters displayed: The higher the number of letters to be recognised, the lower the height of the curve. Therefore, letter report accuracy was poorer in the LVF compared to the RVF because both the drop-off rate of letter legibility with eccentricity and the effect of the number of letters were more pronounced in the LVF.

CP performed very similar to normal readers when letters fell into the LVF. However, considerable differences appeared for performance in the RVF. An ANOVA performed on the patient's letter identification scores showed that letter report accuracy in the RVF dropped drastically with the distance of the letters from fixation [the effect of serial letter position in the RVF was significant for all word lengths, $F(4,16) = 22.86$, $P < .001$; $F(5,20) = 13.88$, $P < .001$; $F(6,24) = 23.42$, $P < .001$; $F(7,28) = 12.48$, $P < .001$; $F(7,28) = 4.96$, $P = .001$ for five- to nine-letter words, respectively); and, contrary to normal participants', performance of the patient decreased significantly with the number of letters presented in the RVF [a significant effect of word length is observed in the RVF; $F(4,16) = 3.36$, $P < .05$]. Thus, the present analysis confirms that CP had a deficit in processing complex visual stimuli presented in his RVF. However, when the data of the patient are displayed in word-centred co-ordinates, CP's difficulties in reporting letters from the second half of words become evident again.

The same letter recognition scores are plotted aligned at the centre of the word for CP (Fig. 11a) and normal participants (Fig. 11b). Thus, in panels I of Figure 11, scores for letters that belong to the first half of words are presented to the left of the central vertical line and scores for letters that belong to the second half of words are presented to the right of this line. Panels II of Fig. 11 give the average of these scores across all viewing conditions. When averaged across all viewing conditions, letter report accuracy for normal participants does not change with the position of letters in the word (first half vs second half) except for five- and six-letter words [$F(1,14) = 19.3$, $P < .001$; $F(1,14) = 8.55$, $P < .05$; $F(1,14) = 1.67$, $P > .1$; $F(1,14) < 1$; $F(1,14) = 1.03$, $P > .1$ for five- to nine-letter words, respectively]. For long words, the serial position functions are slightly U-shaped, indicating that letters at the very beginning and end of the strings were better perceived than letters occupying more central positions. This so called "end-effect" is very well known in vision research and is attributed to low-level visual factors (e.g. Andriessen &

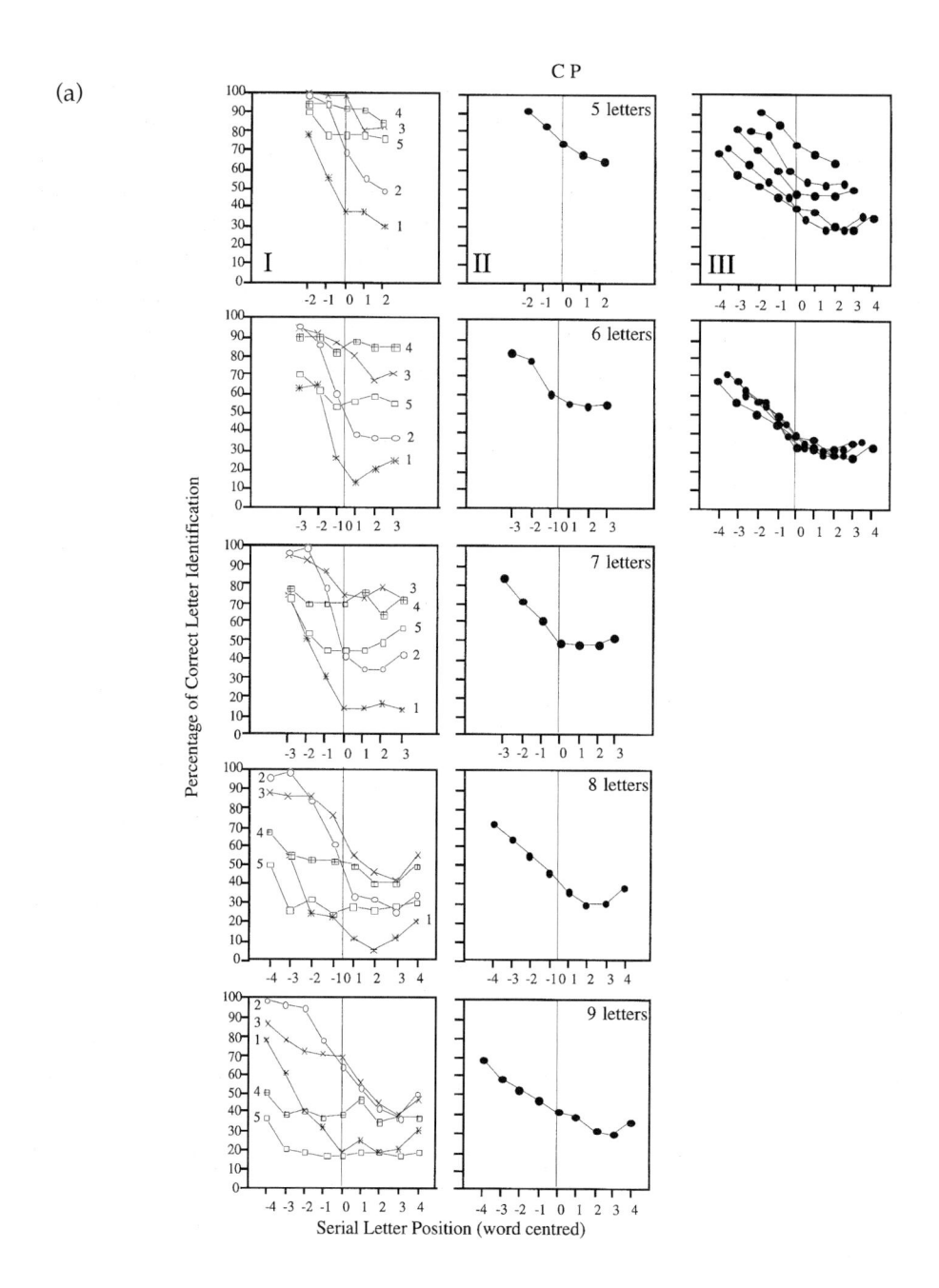

Fig. 11. Percentage of correct letter identification for CP (a) and normal participants (b) as a function of serial letter position (from Experiment 1). Data are aligned at the centre of the word (indicated by the vertical line). Scores for letters that belong to the first half of words are presented to the left of the vertical line, scores for letters that belong to

the second half of words are presented to the right of this line. Panels I give letter identification scores for the different fixation conditions. Panels II give letters identification scores averaged across the fixation conditions. Panels III give the averaged serial-letter position functions for all word lengths (upper panel), equalised in height (lower panel).

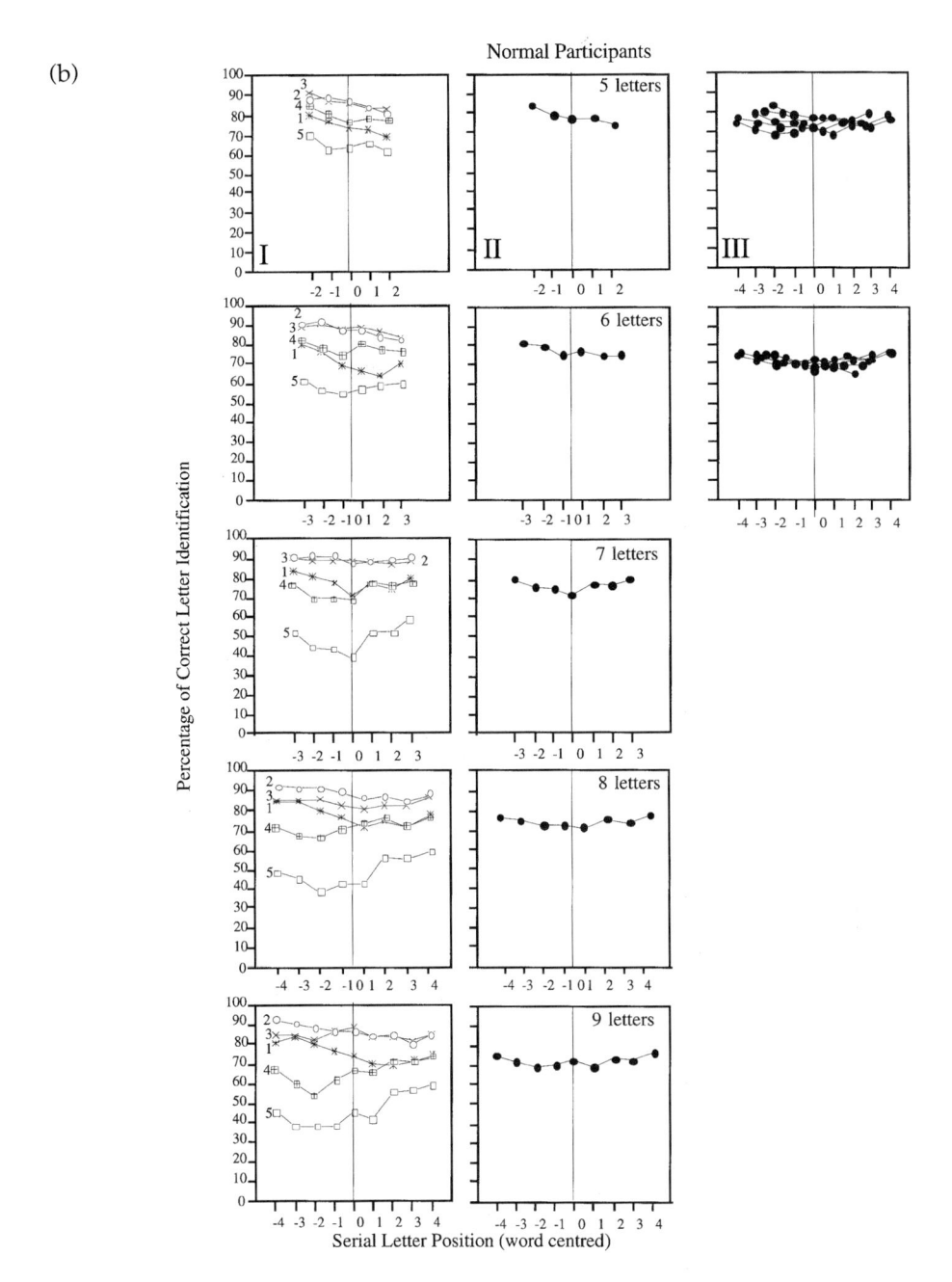

Bouma, 1976; Bouma, 1970). Compared to that of the controls, CP's pattern of performance differs in many aspects (Fig. 11a). An ANOVA indicated that, for all word lengths, letter report accuracy of the patient was significantly higher for letters in the first than in the second half of words [$F(1,4) = 56.94$, $P < .01$; $F(1,4) = 83.9$, $P < .001$; $F(1,4) = 114.9$, $P < .001$; $F(1,4) = 51.84$, $P < .01$; $F(1,4) = 50.08$, $P < .01$ for five- to nine-letter words]. This was true independently of where the word was fixated (see panels I). Averaged over all viewing conditions (panels II), the probability of correct letter report drops linearly from the first to approximately the central letter in the word and remains low in the second half of the word [the interaction between serial letter position in the first and second half of words was significant; $F(2,8) = 45.54$, $P < .001$; $F(2,8) = 44.33$, $P < .001$; $F(3,12) = 71.21$, $P < .001$; $F(3,12) = 12.03$, $P < .001$; $F(4,16) = 34.42$, $P < .001$, for five- to nine-letter words]. Panels III of Fig. 11b give the averaged serial-letter position functions for all word lengths (upper panel), equalised in height (lower panel). The perfect overlap between the curves shows that CP's deviant pattern of performance is better described within a word-centred space than in terms of retinal co-ordinates (see Fig. 10).

Finally, in a last analysis, we looked at letter recognition scores for the letter that appeared at fixation or close to fixation. Because, from trial to trial, words were displayed at different positions relative to the fixation point, the letters that appeared at fixation occupied different positions within the word. Visual resolution for these letters is high and nearly equal, independent of the relative location of these letters in words. If letter report accuracy changes as a function of the position of the fixated letters in words, this might reflect the way letter information is extracted during visual word recognition. The present analysis was carried out only on *incorrect* word responses because, for correct word responses, letter identification scores necessarily mirror word identification scores.

Figure 12 plots viewing-position-dependent letter recognition scores for the letter at or close to fixation. Due to our technique, fixation location was sometimes between two letters. In that case, the average score for these two letters was calculated. The proportion of correct letter identification was obtained by the following calculation: (total letter score – correct word recognition score)/(1 – correct word recognition score). The left panel gives performance for CP, and the right panel gives performance for normal participants. Two points are worth noting. First, for normal participants, report accuracy for letters that appeared at fixation varied with the relative location of letters in words [$F(4,56) = 18.39$, $P < .001$]. The shape of the resulting serial position function is similar to that of the classic viewing position function, even though these letter scores correspond to trials in which words were not correctly reported. Like word recognition accuracy, letter report accuracy was maximum when the word was fixated slightly to the left of its centre (see curve for nine-letter words, for example). Second, CP performed essentially like normal readers, though the variation of letter report accuracy with letter position in words was

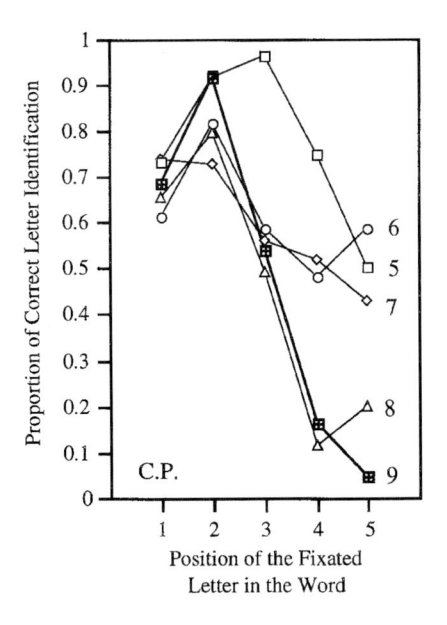

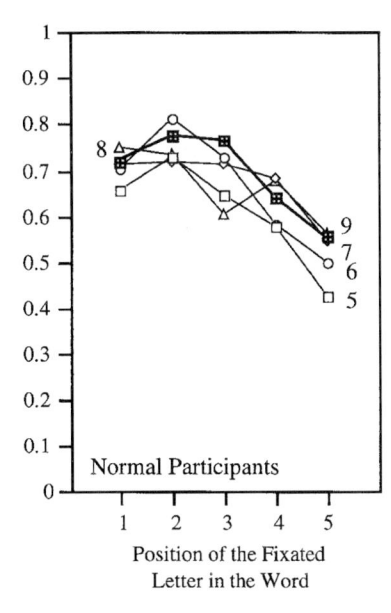

Fig. 12. Proportion of letters near or at fixation correctly identified as a function of letter position in the word, for the different word-lengths (five to nine letters). When fixation was not on a letter but between two letters, the average performance for these two letters was calculated. The left panel gives performance for CP, the right panel gives performance for normal participants. Performance was better for letters in the first half of words than for letters in the second half of words (see thick line for nine-letter words).

much more pronounced for the patient than for normal participants [$F(4,16) = 21.52$, $P < .001$]. Such pattern of performance is surprising given that, contrary to normal readers, CP showed optimal *word* recognition scores when fixating to the right of words' centre. Thus, when lexical activation is not sufficient for correct word identification, similar position-dependent variations in letter report accuracy are observed for normal readers and CP. Note, however, that, in contrast to normal readers, the patient remained very poor in reporting letters placed in the second half of words. The results of the present and previous analyses are therefore coherent with the results obtained in Experiments 3 and 4. The patient seems to have a right hemi-visual field impairment but, more important, he processes letters differently as a function of their relative position in words. A similar position-dependent variation of letter report accuracy was also observed in normal participants, either by experimental manipulation (Experiment 4) or during incorrect word identification (analyses of letter identification scores in Experiment 1).

GENERAL DISCUSSION

Summary

In the present article, we investigate the reading ability of CP, a pure alexic patient, using an experimental paradigm that is known to elicit a highly stable pattern of performance in normal readers. This pattern consists of a systematic variation of reading performance as a function of the position of the eyes in the word: Word recognition performance is best when the eyes fixate slightly left of the centre of the word and decreases when the eyes deviate from this optimal viewing position towards the beginning or towards the end of the letter string (Nazir, 1993; Nazir et al., 1992, in press; O'Regan & Jacobs, 1992; O'Regan et al., 1984). Visual and lexical factors affect the shape and the height respectively of the viewing position curve (McConkie et al., 1989, Nazir et al., 1992, in press; O'Regan & Jacobs, 1992; Vitu, 1991). A mathematical modal (Nazir et al., 1991), which provides a good description and quantification of the viewing position effect, served to interpret deviations from the norm of CP's reading pattern.

Like normal readers, CP showed a systematic variation of reading performance when his eyes were fixating different locations in the word (Experiment 1), suggesting that the system that supports normal reading was at least partly functioning. However, the shape of this viewing position function was remarkably different from the norm. Whereas normal readers showed the typical inverted U-shape function with a maximum slightly to the left of word centre, CP showed an inverted U-shape function with a maximum to the right of word centre. This result suggests that CP had difficulties in processing letters presented to the right of fixation. Further investigations revealed two sources that underlie these difficulties: (1) a general visual impairment in processing complex stimuli presented in the right visual field (Experiment 2), and (2) poor processing of letters placed towards the end of words, independent of the visual field concerned (Experiments 3 and 4). The latter was confirmed by additional analyses performed on the data of Experiment 1.

Simulation of CP's Viewing Position Effect

Within the framework of Nazir et al.'s model, these two deficits are sufficient to account for CP's unusual viewing position effect. To simulate the impact of a general visual impairment in processing complex stimuli presented in the right visual field, we decreased the values of all entries on the right side of the diagonal in Table 1 by a constant value of either .05, .10, .20, or .30 (see Appendix E for more details). A higher value represents a stronger visual "impairment". The left panel in Fig. 13 gives the resulting transition of the viewing position effect from the normal quadratic function towards a linear function as the "impairment" becomes stronger. Note that none of these functions captures the strong drop-off of CP's reading performance when he moved fixation from the fourth to the fifth fixation position. Thus, a pure right visual field defect cannot fully account for CP's abnormal viewing position effect. To simulate a deficit in processing letters of the second half of words, we decreased the recognition probabilities of the third, fourth, and fifth letter of a five-letter word, except when these letters were directly fixated or adjacent to the directly fixated letter. Probabilities were decreased by a constant value of either .20, .30, or .40 (see Appendix F for more details). When the deficit in processing letters in the second half of words is added to the general visual impairment that involves the right visual field, the resulting viewing position curves mimic the basic pattern of CP's viewing position effect (see circles in the right panel of Fig. 13). Hence, the deficit in processing letters placed at the right side of words seems to be an important factor in determining CP's unusual viewing position effect.

General Interpretation

By implicitly manipulating participants' response strategies and/or the information they may use for responding, we could also observe variations in the processing of the component letters of words in normal readers. When lexical information could be used to identify a target letter without completing the visual analysis of the letter string, normal readers performed equally well on all letter positions in the word (Experiment 3). However, when participants were constrained to exploit letter information fully to discriminate words from pseudowords, normal readers mimicked CP's pattern of performance: They performed poorly when the critical letter of a pseudoword was placed towards the end of the string (Experiment 4). In addition, on normal participants' incorrect word responses in Experiment 1, report scores for letters at fixation were higher for letters in the first half of words than for letters in the second half of words (cf. Further Analyses section). Thus, it seems that the availability of letter information differs generally as a function of the relative spatial location of the letter in the word. Under normal reading conditions, however, this difference might be masked by additional lexical or contextual information. CP's reduced ability to report letters placed in the second half of words is therefore not necessarily the cause of his reading deficit. Rather, the real impairment could consist of a dysfunction of the coupling, between incoming visual information and stored lexical knowledge, that is hypothesised to underlie normal reading (e.g. McClelland & Rumelhart, 1981). Thus, according to this view, this dysfunction uncovers the existence of a prelexical stage of word processing in which letter information is weighted differently as a function of the position of the letter in a word-centred space. Arguments in favour of this interpretation are developed next.

Coding the Position of Letters in Words

The existence of a stage of word processing where the position of letters in words is coded has been discussed earlier by other investigators (Caramazza & Hillis, 1990; Humphreys, Evett, & Quinlan, 1990; Peressotti & Grainger, 1995; Rapp & Caramazza, 1991). Humphreys et al. (1990), for instance, have shown that primes facilitate the identification of target words, provided that prime and target pairs share letters at the same relative position in the string. For example, the prime "btvuk" facilitates the recognition of the target word "BLACK" whereas the prime "tbvku" fails to

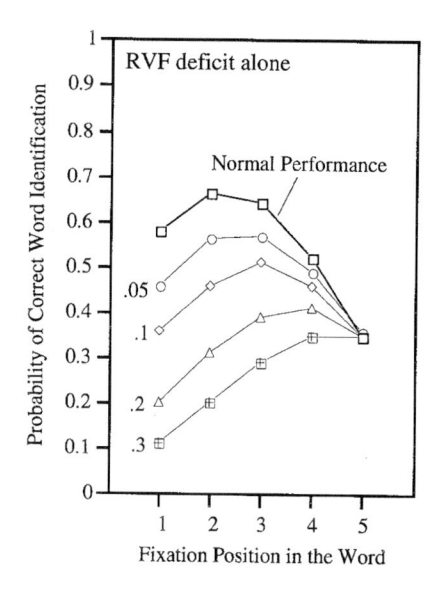

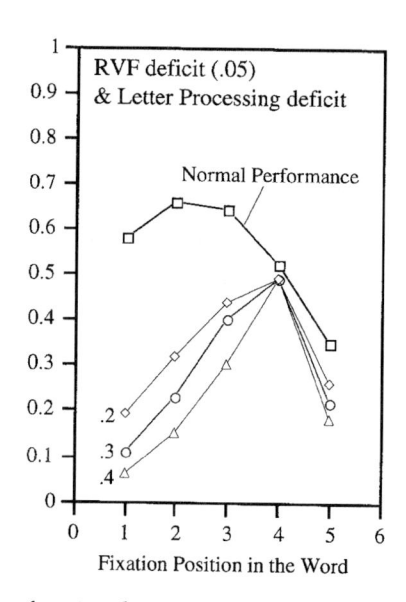

Fig. 13. Theoretical viewing position curves predicted by the model for a five-letter string. The left panel shows the predicted curves when the probability of letter identification in the right visual field is reduced by a value of either .05, .10, .20 or .30. The right panel shows the predicted curves when, in addition to a right visual field impairment of .05, the probability of identifying letters placed in the second part of the string is decreased by a value of either .20, .30, or .40 (see Appendix F for more details).

produce any facilitation. Similarly, Grainger and Jacobs (1991) have shown that when a target letter is embedded in a string of hash marks (e.g. ##T##), priming effects are obtained if the position of the target letter in a prime word is identical to that in the target string (e.g. table / T####), but no effect is observed if the positions differ (e.g. table / ####T). According to Caramazza and Hillis (1990), visual processes involved in word recognition can be described in terms of three stages of analysis: (1) the computation of a retinocentric feature map; (2) the computation of a stimulus-centred letter-shape map; and (3) the computation of a word-centred grapheme description (inspired by Marr, 1982). Whereas, at the first level, information is computed in a retinocentric space, at the second and third levels, information is computed in a co-ordinate system centred on the stimulus. The second level codes the shape and spatial relations of letters. The third level computes information about abstract letter identities and their relative position in the word. CP's pattern of reading performance is compatible with the existence of such levels of processing, where spatial information about letters is coded in a stimulus- or word-centred space, because his difficulties in processing letters were restricted to the second half of words, independent of conditions of presentation (Experiments 3 and 4 as well additional analyses of Experiment 1). Caramazza and colleagues found similar spatially defined patterns of impaired reading in a neglect dyslexic patient and in a pure alexic patient (Caramazza & Hillis, 1990; Rapp & Caramazza, 1991).

Weighted Letter Positions

We claim that, at the stage of word processing where the relative position of letters is coded, the availability of letter information varies with the relative location of the letter in the string. In support of this claim, we have shown in Experiment 4 that, when normal participants had to exploit letter information fully to discriminate words from pseudowords, their performance varied with the position of the critical letter in the string. In the same vein, the analyses at the letter level performed on the data of Experiment 1 showed that, for normal participants as well as for the patient, identification scores of letters at fixation varied with the relative position of these letters in the word, although visual resolution for the fixated letters was high and independent of letter position in the word. The shape of the letter identification function was quite similar to the classic viewing position curve observed in word identification tasks. Nazir and Montant (1998) found similar results in a recent study. Letter identification scores in a two-alternative forced-choice task varied with the relative position of target letters in words, although the target letter was displayed at fixation location: Performance was maximal for letters placed towards the left of the words' centre and decreased for letters in the right half of words. Together, these results suggest that, in the normal reading system, the availability of information from letters differs as a function of their relative position in words. According to Nazir (Nazir et al., in press; Nazir, in press), this position-dependent variation in letter coding could result from perceptual learning. During normal reading, most words are identified in one single fixation. Given the structure of the retina, visual cues available from the letters of a word differ as a function of the position of these letters relative to the centre of gaze. The closer the letter to fixation, the more details about the letter are available. By contrast, for parafoveally presented letters, the available visual information is coarse and ambiguous. Given that during reading the eyes tend to land slightly to the left of the word's centre (e.g. McConkie, Kerr, Reddix, & Zola, 1988; Rayner, 1979; Vitu, O'Regan, & Mittau, 1990), the reading system is trained to perceive words with high resolution for letters in the first half of the word and poor resolution for letters placed towards the end of the word. With experience, the reading system weights the letter units as a function of the quality of the available visual information.

The Coupling between Pre-lexical and Lexical Information

Considering reading as an interactive process (McClelland & Rumelhart, 1981), we argue that a dysfunction of the coupling between pre-lexical and lexical information is responsible for CP's reading difficulties (see Behrmann et al., this issue, for a similar view). This dysfunction uncovers a stage of word processing where letter information is weighted differently as a function of the relative location of the letter in the string. Under normal reading conditions, this difference in the availability of letter information is masked by additional activation due to lexical or contextual information. In support of this, we have shown that letters belonging to the second half of words are more difficult to process either when the reading system is impaired, as in CP, or, as was shown for normal participants, when lexical activation is experimentally reduced (Experiment 4) or not sufficient for correct word identification (analysis at the letter level of the data of Experiment 1). To strengthen this interpretation, we simulated the consequences of a disconnection between letter units and word units in the Interactive Activation Model (IAM; McClelland & Rumelhart, 1981). The model contains three levels of processing units (letter-feature units, feature units, and word units) with feedforward and feedback connections. In the model, consistent elements mutually support each other (e.g. the letter U in the third position and the word BLUE) while inconsistent or competing elements mutually inhibit each other (e.g. the words BLUE and BLUR). For the simulations, we used an extension of the original architecture and parameters of the IAM to five-letter words (Grainger & Jacobs, 1994; Jacobs & Grainger, 1992). Twenty English 5-letter words served as stimuli: 10 high-frequency words (mean frequency 845.4 occurrences per million) and 10 low-frequency words (mean frequency 5.6 occurrences per million). Letter units were weighted differently as a function of the relative position of each letter in the string (instead of the original unique value of 1, the strength of the letter units was set to .9, 1, .95, .8 and .7 for the first to the fifth letter position respectively in a five-letter word). The dysfunction of the coupling between prelexical and lexical information was simulated by interrupting the feedback excitatory connections between the letter and word levels. The results of the simulations are given in Fig. 14 and 15.

Figure 14 (squares) shows that, when the architecture of the model is intact, activation in the letter units hardly vary with the strength of these units at each position in the word. Consistent with our hypothesis, this position-dependent variation of activation in the letter units dramatically increases when feedback connections in the model are interrupted (see circles in Fig. 14). This result suggests that, under normal reading conditions, the feedback flow of lexical activation masks the structural properties of a prelexical level of processing. The simulations also reveal that the frequency effect in the word units is stronger when the feedback connections between word and letter levels are interrupted (see right panel of Fig. 15) than when these connections are intact (see left panel of Fig. 15). This corroborates the results of the posthoc analyses performed on the data of Experiment 1: CP had a frequency effect (21.1%) comparable or even larger in size to that of normal participants (15.63%), suggesting that, during visual word recognition, lexical knowledge was at least as much activated in CP as in normal participants. Note that frequency effects, as well as other lexical effects, were found in several pure alexic patients over the last decade (see Behrmann et al, this issue, for a review).

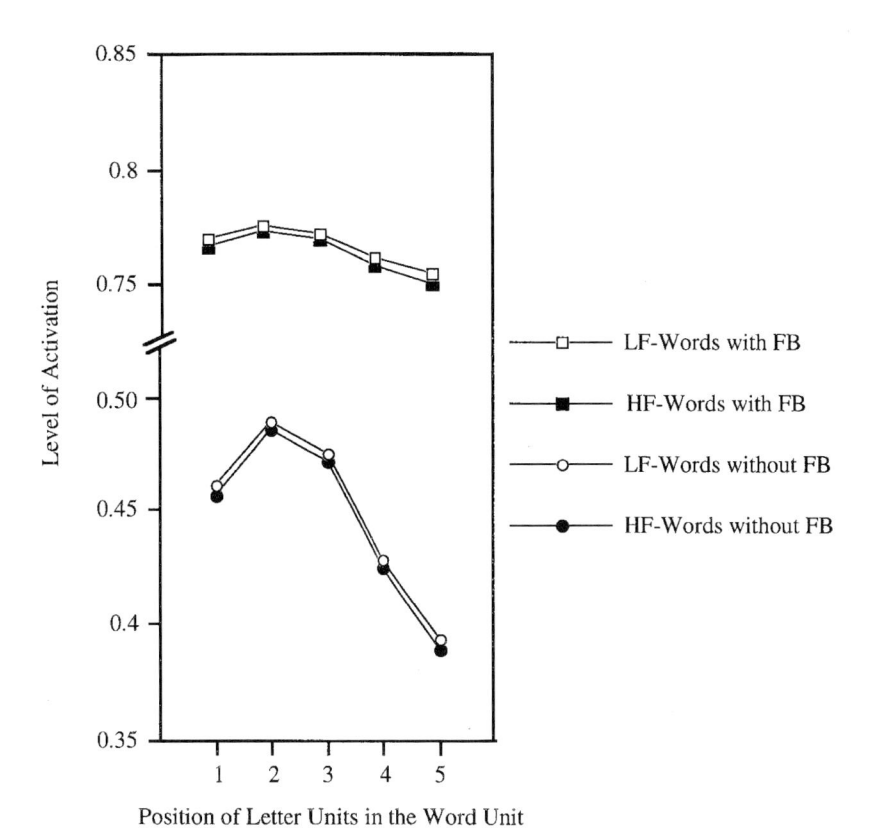

Fig. 14. Level of activation in the letter units of the IAM for high-frequency and low-frequency words as a function of the relative position of letters in the word. Activation in the letter units is observed after 40 cycles. Upper curves give the activation in the letter units when word-to-letter feedback connections are intact. Lower curves give the activation in the letter units when word-to-letter feedback connections are disrupted.

The Coupling and the Length Effect

Finally, the hallmark of pure alexia—the word length effect—could also be accounted for in terms of a dysfunction in the prelexical/lexical interactions. Given the strong drop of acuity in parafoveal vision, it is rather puzzling that adult readers do not show a marked word length effect during reading. In fact, as demonstrated in Fig. 16, the model by Nazir et al. (1991) predicts a systematic decrease of performance with every additional letter in the word.

The theoretically predicted length effect is actually present in beginning readers until they have 3 or 4 years of reading instruction (Aghababian & Nazir, 1998). Although children already show the classic viewing position effect after few weeks of training, many more years are needed before a six-letter word can be read as efficiently as a four-letter word. Therefore, it seems that the absence of a length effect in adults is the result of an adaptation

Bub, D.N., & Arguin, M. (1995). Visual word activation in pure alexia. *Brain and Language, 49*, 77–103.

Bub, D.N., Black, S.E., & Howell, J. (1989). Word recognition and orthographic context. *Brain and Language, 37*, 357–376.

Caramazza, A., & Hillis, A.E. (1990). Levels of representation, co-ordinate frames, and unilateral neglect. *Cognitive Neuropsychology, 7*, 391–445.

Carr, T.H., & Posner, M.I. (1994). The impact of learning to read on the functional anatomy of language processing. In B. de Gelder & J. Morais (Eds.), *Language and literacy: Comparative approaches*, Cambridge, MA: MIT Press.

Coltheart, M., Davelaar, E., Jonasson, J.T., & Besner, D. (1977). Access to the internal lexicon. In S. Dornic (Ed.), *Attention and performance VI*. London: Academic Press.

Content, A., Mousty, P., & Radeau, M. (1990). Brulex, une base de données lexicales informatisées pour le français écrit et parlé. *L'Année Psychologique, 90*, 551–566.

Coslett, H.B., & Saffran, E.M. (1989). Evidence for preserved reading in "pure" alexia. *Brain, 112, 327–359.*

Coslett, H.B., Saffran, E.M., Greenbaum, S., & Schwartz, H. (1993). Reading in pure alexia: The effect of strategy. *Brain, 116*, 21–37.

Crovitz, H.F., & Zener, K. (1962). A group-test for assessing hand- and eye-dominance. *American Journal of Psychology, 75*, 271–276.

Déjerine, J. (1882). Contributions à l'étude anatomopathologique et clinique des différentes variétés de cécité verbale. *Mémoires de la Société Biologique, 44*, 61–90.

Farah, M.J., & Wallace, M. (1991). Pure alexia as a visual impairment: A reconsideration. *Cognitive Neuropsychology, 3*, 149–177.

Farid, M., & Grainger, J. (1996). How initial fixation position influences visual word recognition: A comparison of French and Arabic. *Brain and Language, 53*, 351–368.

Friedman, R.B., & Alexander, M.P. (1984). Pictures, images, and pure alexia: A case study. *Cognitive Neuropsychology, 1*, 9–23.

Grainger, J., & Jacobs, A.M. (1991). Masked constituent letter priming in an alphabetic decision task. *European Journal of Cognitive Psychology, 3*, 413–434.

Grainger, J., & Jacobs, A.M. (1994). A dual read-out model of word context effects in letter perception: Further investigations of the word superiority effect. *Journal of Experimental Psychology: Human Perception and Performance, 20*, 1158–1176.

Grainger, J., & Jacobs, A.M. (1996). Orthographic processing in visual word recognition: A multiple read-out model. *Psychological Review, 103*, 518–565.

Grier, J.B. (1971). Nonparametric indices for sensitivity and bias: Computing formulas. *Psychological Bulletin, 75*, 424–429.

Hagenzieker, M.P., van der Heijden, A.H.C., & Hagenaar, R. (1990). The time courses in visual information processing: Some empirical evidence for inhibition. *Psychological Research, 52*, 13–21.

Hanley, J., & Kay, R. (1992). Does letter-by-letter reading involve the spelling system? *Neuropsychologia, 30*, 237–256.

Howard, D. (1991). Letter-by-letter readers: Evidence for parallel processing. In D. Besner & G.W. Humphreys (Eds.), *Basic processes in reading: Visual word recognition*. Hove, UK: Lawrence Erlbaum Associates Ltd.

Humphreys, G.W., Evett, L.J., & Quinlan, P.T. (1990). Orthographic processing in visual word identification. *Cognitive Psychology, 22*, 517–560.

Jacobs, A.M., & Grainger, J. (1992). Testing a semistochastic variant of the interactive activation model in different word recognition experiments. *Journal of Experimental Psychology: Human Perception and Performance, 18*, 1174–1188.

Kay, R, & Hanley, J. (1991). Simultaneous form perception and serial letter recognition in a case of letter-by-letter reading. *Cognitive Neuropsychology, 8*, 249–273.

Kinsbourne, M., & Warrington, E.K. (1962). A disorder of simultaneous form perception. *Brain, 85*, 461–486.

Levine, D.M., & Calvanio, R. (1978). A study of the visual defect in verbal alexia—simultagnosia. *Brain, 101*, 65–81.

Loftus, G.R., & Masson, M.E.J. (1994). Using confidence intervals in within-subject designs. *Psychonomic Bulletin and Review, 1*, 476–490.

Marr, D. (1982). Vision, New York: W.H. Freeman & Co.

Massaro, D.W., & Klitzke, D. (1977). Letters are functional in word identification. *Memory and Cognition, 5*, 292–298.

McClelland, J.L. (1976). Preliminary letter identification in the perception of words and nonwords. *Journal of Experimental Psychology: Human Perception and Performance, 2*, 80–91.

McClelland, J.L., & Rumelhart, D.E. (1981). An interactive activation model of context effect in letter perception. Part I: An account of basic findings. *Psychological Review, 88*, 375–407.

McConkie, G.W., Kerr, P.W., Reddix, M.D., & Zola, D. (1988). Eye movement control during reading: I. The location of initial eye fixations on words. *Vision Research, 28*, 1107–1118.

McConkie, G.W., Kerr, P.W., Reddix, M.D., Zola, D., & Jacobs, A.M. (1989). Eye movement control during reading: II. Frequency of refixating a word. *Perception and Psychophysics, 46*, 245–253.

Montant, M., & Nazir, T.A. (1995). *Coding of spatial information of visually presented words*. Proceedings of the 36th Annual Meeting of the Psychonomic Society, Los Angeles, 563.

Nazir, T.A. (1993). On the relation between the optimal and the preferred viewing position in words during reading. In G. Ydewalle & J. van Rensbergen (Eds.), *Perception and cognition: Advances in eye movement research* (pp. 349–361). Amsterdam: North-Holland.

Nazir, T.A. (in press). Les mouvements oculaires et la lecture. In Boucard, Heniff, & Belin (Eds.), *Vision = aspects perceptifs et cognitifs*. Marseilles: Editions SOLAL.

Nazir, T.A., Heller, D., & Sussmann, C. (1992). Letter visibility and word recognition: The optimal viewing position in printed words. *Perception and Psychophysics, 52*, 315–328.

Nazir, T.A., Jacobs, A.M., & O'Regan, J.K. (in press). Letter legibility and visual word recognition. *Memory and Cognition*.

Nazir, T.A., & Montant, M. (1998). *The availability of letter information in words and consonant strings*. Manuscript in preparation.

Nazir, T.A., & O'Regan, J.K., & Jacobs, A.M. (1991). On words and their letters. *Bulletin of the Psychonomic Society, 29*, 171–174.

Olzak, L.A., & Thomas, J.P. (1986). Seeing spatial patterns. In K.R. Boff, L. Kaufman, & J.P. Thomas (Eds.), *Handbook of perception and human performance, Vol, II*, (pp. 7:1–7:56). New York: Wiley.

O'Regan, J.K., & Jacobs, A.M. (1992). Optimal viewing position effect in word recognition: A challenge to current theory. *Journal of Experimental Psychology: Human Perception and Performance, 18*, 185–197.

O'Regan, J.K., Lévy-Schoen, A., Pynte, J., & Brugaillère, B. (1984). Conveniente fixation location within isolated words of different length and structure. *Journal of Experimental Psychology: Human Perception and Performance, 10*, 250–257.

Paap, K.R., & Johansen, L.S. (1994). The case of the vanishing frequency effect: A retest of the verification model. *Journal of Experimental Psychology: Human Perception and Performance, 20*, 1129–1157.

Patterson, K., & Kay, J. (1982). Letter-by-letter reading: Psychological descriptions of a neurological syndrome. *Quarterly Journal of Experimental Psychology, 34A*, 411–441.

Peressotti, F., & Grainger, J. (1995). Letter position coding in random consonant arrays. *Perception and Psychophysics, 57*, 875–890.

Posner, M.I. & Carr, T.H. (1992). Lexical access and the brain: Anatomical constraints on cognitive models of word recognition. *American Journal of Psychology, 105*, 1–26.

Rapp, B.C., & Caramazza, A. (1991). Spatially determined deficits in letter and word processing. *Cognitive Neuropsychology, 8*, 275–311.

Rayner, K. (1979). Eye guidance in reading: Fixation locations within words. *Perception, 8*, 21–30.

Reicher, G.M. (1969). Perceptual recognition as a function of meaningfulness of stimulus material. *Journal of Experimental Psychology, 81*, 275–281.

Reuter-Lorenz, P.A., & Brunn, J.L. (1990). A prelexical basis for letter-by-letter reading: A case study. *Cognitive Neuropsychology, 7*, 1–20.

Rumelhart, D.E., & McClelland, J.L. (1982). An interactive activation model of context effects in letter perception: Part II. The contextual enhancement and some tests and extensions of the model. *Psychological Review, 89*, 60–94.

Saffran, E.M., & Coslett, H.B. (this issue). Implicit vs. letter-by-letter reading in pure alexia: A tale of two systems. *Cognitive Neuropsychology, 15*(1/2).

Sekuler, E.B., & Behrmann, M. (1996). Perceptual cues in pure alexia. *Cognitive Neuropsychology, 13*, 941–974.

Shallice, T., & Saffran, E. (1986). Lexical processing in the absence of explicit word identification: Evidence from a letter-by-letter. *Cognitive Neuropsychology, 3*, 429–458.

Vitu, F. (1991). The influence of reading rhythm on the optimal landing position effect. *Perception and Psychophysics, 50*, 58–75.

Vitu, F., O'Reagan, J.K., & Mittau, M. (1990). Optimal landing position in reading isolated words and continuous texts. *Perception and Psychophysics, 47*, 583–600.

Warrington, E.K., & Shallice, T. (1980). Word-form dyslexia. *Brain, 103*, 99–112.

Ziegler, J.C., & Jacobs, A.M. (1995). Phonological information provides early sources of constraint in the processing of letter strings. *Journal of Memory and Language, 34*, 567–593.

Ziegler, J.C., Van Orden, G.C., & Jacobs, A.M. (1997). Phonology can help or hurt the perception of print. *Journal of Experimental Psychology: Human Perception and Performance, 23*, 845–860.

APPENDIX A

List of the Stimuli Used in Experiment 1

"Pos" means fixation position in the word; "Freq" means frequency of occurrence (× per 100 millions)

Pos	5-letter Words	Freq	6-letter Words	Freq	7-letter Words	Freq
1	AVARE	1744	COFFRE	1029	GRAVURE	1029
1	IDOLE	1688	AVERSE	1038	EVASION	1072
1	PLOMB	1408	PIQURE	1084	DORMANT	1131
1	REPLI	1361	CERTES	11507	TORRENT	2054
1	HOTEL	12954	IRREEL	1135	SERIEUX	10729
1	TUYAU	1021	ASPECT	10252	PLACARD	1080
1	GUERI	1842	RIGIDE	1318	PELERIN	1135
1	ETANG	1740	NUDITE	1029	VITRINE	1157
1	ASSIS	14239	COCHON	1791	SYSTEM	11639
1	DEVOT	1408	IVOIRE	1067	EDITEUR	1063
2	GARNI	1382	ADULTE	1939	ECOLIER	1106
2	PETIT	147638	GENANT	1106	CORDIAL	1191
2	OFFRE	1391	JARDIN	14975	MALHEUR	11290
2	ECUME	1327	PAQUES	1157	BARBARE	1791
2	ACHAT	1148	ENCLOS	1233	DENSITE	1059
2	BIJOU	1671	SERMON	1250	CABARET	1067
2	ENVIE	11507	PROJET	11061	LYRISME	1106

Pos	Words	Freq	Words	Freq	Words	Freq
2	GENOU	9227	VISUEL	1761	THEATRE	11452
2	PITIE	8002	BUFFET	1050	PLANETE	2029
2	RIVAL	1565	CIRQUE	1140	DOUTEUX	1148
3	TALUS	1378	ULTIME	1152	ENTENTE	2080
3	ODEUR	12214	EPERDU	1284	SORDIDE	1131
3	AMPLE	1208	PREFET	1033	SPATIAL	1152
3	QUETE	1250	DEESSE	1072	JUSTICE	10125
3	APPEL	9406	SIRENE	1084	PARFAIT	10963
3	TYRAN	1514	POESIE	10720	GRAISSE	1084
3	GENIE	10657	DIFFUS	1110	CLAMEUR	1127
3	SOUCI	8291	DEFUNT	1135	ETOURDI	1067
3	EXCLU	1127	AMBIGU	1267	AUBERGE	2025
3	INDEX	1557	DEBOUT	11048	SOUHAIT	1097
4	ELOGE	1897	MARAIS	1101	AUSTERE	1144
4	IMPUR	1067	FURTIF	1123	REPONSE	11805
4	JETEE	1846	DEPENS	1114	LIMPIDE	1191
4	SCENE	13601	TARDIF	1174	DOCTEUR	11205
4	FOUTU	1046	MANUEL	1280	ACCUEIL	1778
4	ROMAN	11337	SOLDAT	11167	IVROGNE	1097
4	IMPIE	1130	ECURIE	1182	CRUAUTE	2025
4	OPERA	1671	ACCORD	11082	CORRECT	1174
4	IMPOT	1012	CRAYON	1939	ANORMAL	1195
4	INOUI	1450	LEGUME	1165	FUNESTE	1050
5	TREVE	1059	CONVOI	1169	PRODIGE	1110
5	TRIBU	1340	BLOUSE	1220	COLOMBE	1050
5	GENRE	12571	MOULIN	1816	ARTISAN	1212
5	BOURG	1850	BLAGUE	1182	SOMMEIL	10269
5	GRIEF	1225	DOLLAR	1246	COMPACT	1063
5	CHAOS	1527	PERRON	1284	INCONNU	11614
5	EVEIL	1110	DRAGON	1029	ARBITRE	1131
5	MAMAN	12273	ORDURE	1089	BRUYANT	1157
5	PAIEN	1233	MADAME	10946	PROPICE	1033
5	DEGRE	8427	DIVERS	11248	JESUITE	2016

	8-letter		9-letter	
Pos	Words	Freq	Words	Freq
1	EXCITANT	1038	ASCENSION	1646
1	BIOLOGIE	1046	COURAGEUX	1650
1	INSTINCT	10486	ENTOURAGE	1259
1	SEPTIEME	1063	ETINCELLE	1340
1	SYMPTOME	1080	CONDITION	21756
1	SOMMAIRE	1093	FRATERNAL	1446
1	SOLIDITE	1114	EVENEMENT	13847
1	MAINTIEN	1135	CLIENTELE	1361
1	TERRIBLE	10737	ADORATION	1208
1	CLINIQUE	1038	BRUTALITE	1118

2	NOURRICE	1059	CAUSALITE	1318
2	INSTABLE	1076	DECHIRANT	1374
2	CRITIQUE	10167	ECLAIRAGE	1425
2	FIEVREUX	1157	HONNETETE	1220
2	ANGOISSE	10448	CARACTERE	20084
2	MARITIME	1033	DYNAMIQUE	1267
2	CORRIDOR	1046	ABSURDITE	1391
2	AUDIENCE	1063	ATTENTION	16723
2	MILLIARD	1072	CHAUSSURE	1425
2	EVENTAIL	1080	ENERGIQUE	1471
3	EPHEMERE	1097	FINANCIER	1642
3	ANECDOTE	1123	CARREFOUR	1471
3	CHAPELET	1148	ACCOUTUME	1323
3	FAISCEAU	1101	CASQUETTE	1484
3	TYRANNIE	1033	ESCLAVAGE	1471
3	JEUNESSE	11001	ESSENTIEL	10048
3	AUTONOME	1123	GRATITUDE	1352
3	CAMARADE	10682	BLANCHEUR	1097
3	BIENFAIT	1055	DIFFICILE	16825
3	DOUZAINE	1063	CARESSANT	1161
4	PEUPLIER	1080	DERISOIRE	1425
4	IMMEUBLE	1135	DIRIGEANT	1178
4	DIVISION	10329	CHANGEAT	1552
4	MONOPOLE	1059	DAVANTAGE	12750
4	ACTIVITE	10967	EMPREINTE	1352
4	METAIRIE	1076	DEMEURANT	1208
4	ILLIMITE	1076	DIRECTION	10520
4	FEUILLET	1131	FORCEMENT	1620
4	ARAIGNEE	1042	HONORABLE	1242
4	EFFUSION	1046	FRANCHISE	1680
5	REPOUSSE	1063	AMBITIEUX	1369
5	BANQUIER	1063	AUDACIEUX	1123
5	MEDIEVAL	1097	EXTERIEUR	11924
5	INDIVIDU	10746	CHATIMENT	1131
5	ADORABLE	1055	CHRONIQUE	1097
5	INFERNAL	1080	EXPANSION	1314
5	MONTAGNE	10129	AUTREFOIS	10610
5	ATTRIBUT	1072	DISSIMULE	1084
5	AGRICOLE	1089	ELEVATION	1318
5	ENCIENTE	1144	EXTENSION	1365

APPENDIX B

List of the Stimuli Used in Experiment 2

Yes	Trial	No	Trial	Yes	Trial	No	Trial
B	B	B	J	H	H	H	R
B	b	B	j	H	h	H	r
b	b	b	j	h	h	h	r
D	D	D	F	Q	Q	Q	N
D	d	D	f	Q	q	Q	n
d	d	d	f	q	q	q	n
G	G	G	L	T	T	T	M
G	g	G	l	T	t	T	m
g	g	g	l	t	t	t	m

APPENDIX C

List of the Stimuli Used in Experiment 3

"Freq" means frequency of occurrence (× per 100 millions)

Stimulus	Freq	Gaze Location	Target Letter	Response Alternative	Target Position
danse	4228	1	D	F	1
écran	1523	1	E	I	1
enfin	51860	1	E	A	1
image	28325	1	I	U	1
laine	2480	1	L	P	1
nuage	6411	1	N	D	1
rouge	18855	1	R	L	1
salle	15396	1	S	F	1
tapis	3458	1	T	V	1
usage	7947	1	U	O	1
idole	1688	1	D	H	2
jeudi	2484	1	E	O	2
rendu	9631	1	E	O	2
vigne	2790	1	I	A	2
glace	5900	1	L	P	2
envoi	1131	1	N	S	2
brûlé	2561	1	R	I	2
usine	3977	1	S	B	2
stade	1629	1	T	L	2

juste	20833	1	U	A	2
ordre	38003	1	D	F	3
boeuf	2650	1	E	C	3
mieux	49129	1	E	A	3
boire	12269	1	I	A	3
belge	2897	1	L	R	3
bande	4194	1	N	U	3
barbe	4203	1	R	T	3
geste	24232	1	S	L	3
actif	4841	1	T	D	3
jaune	6500	1	U	I	3
garde	11290	1	D	F	4
raser	1063	1	E	U	4
voter	1059	1	E	A	4
clair	13741	1	I	U	4
frèle	1093	1	L	T	4
front	19782	1	N	I	4
heure	86928	1	R	L	4
torse	1207	1	S	P	4
photo	1561	1	T	B	4
époux	2050	1	U	I	4
blond	2003	1	D	T	5
arbre	20625	1	E	I	5
copie	1408	1	E	U	5
mardi	1995	1	I	E	5
moral	20348	1	L	T	5
chien	12112	1	N	L	5
coeur	60462	1	R	S	5
cours	14719	1	S	A	5
bruit	22662	1	T	S	5
noyau	1582	1	U	I	5
désir	24441	3	D	B	1
ennui	6636	3	E	A	1
épais	6007	3	E	U	1
issue	3505	3	I	A	1
libre	21335	3	L	J	1
neige	6585	3	N	D	1
riche	10980	3	R	T	1
siège	4062	3	S	V	1
terre	45134	3	T	L	1
utile	5743	3	U	O	1
idéal	5777	3	D	V	2
hélas	6875	3	E	U	2
métro	1454	3	E	I	2
aimer	76646	3	I	U	2
élite	1637	3	L	M	2
unité	13677	3	N	L	2

grave	12069	3	R	U	2
asile	1608	3	S	R	2
style	6964	3	T	P	2
fumer	4160	3	U	O	2
radio	2390	3	D	N	3
lueur	4837	3	E	A	3
plein	37986	3	E	U	3
doigt	16655	3	I	A	3
folie	6594	3	L	D	3
finir	19459	3	N	L	3
corps	54417	3	R	D	3
fusil	4467	3	S	J	3
coton	1118	3	T	B	3
lourd	13873	3	U	I	3
poids	7313	3	D	T	4
forêt	7823	3	E	I	4
poser	19289	3	E	I	4
motif	5853	3	I	U	4
fable	1386	3	L	R	4
veine	2246	3	N	R	4
écart	2837	3	R	S	4
pause	2301	3	S	F	4
puits	2372	3	T	L	4
celui	69639	3	U	O	4
froid	16715	3	D	S	5
livre	37080	3	E	O	5
vache	3526	3	E	U	5
parmi	20425	3	I	E	5
outil	2514	3	L	S	5
rayon	5875	3	N	L	5
tenir	70966	3	R	D	5
héros	5926	3	S	T	5
géant	2339	3	T	S	5
tordu	1140	3	U	E	5
doute	4398	5	D	P	1
écrit	19824	5	E	A	1
exact	5722	5	E	I	1
icône	144	5	I	A	1
loger	2101	5	L	V	1
nuque	2595	5	N	C	1
règle	9078	5	R	P	1
salut	6500	5	S	N	1
tuile	1284	5	T	C	1
union	4769	5	U	E	1
idiot	3696	5	D	F	2
refus	4016	5	E	O	2
venir	1E + 05	5	E	A	2

sitôt	2922	5	I	U	2
globe	1548	5	L	R	2
entre	93254	5	N	L	2
frais	2807	5	R	L	2
astre	2424	5	S	I	2
étage	2816	5	T	F	2
suite	31005	5	U	O	2
vider	2939	5	D	T	3
agent	4335	5	E	I	3
sueur	3441	5	E	O	3
épine	1242	5	I	U	3
salon	9623	5	L	F	3
vingt	20841	5	N	S	3
herbe	8572	5	R	S	3
liste	1752	5	S	F	3
matin	35885	5	T	D	3
poule	2182	5	U	I	3
coude	3118	5	D	T	4
cruel	4760	5	E	I	4
filet	2382	5	E	A	4
grain	3926	5	I	U	4
perle	2148	5	L	V	4
chêne	2807	5	N	D	4
foire	1752	5	R	S	4
ainsi	81721	5	S	R	4
chute	1028	5	T	P	4
orgue	1186	5	U	I	4
sourd	5870	5	D	T	5
fille	43921	5	E	U	5
proie	3875	5	E	S	5
paroi	2067	5	I	U	5
total	8210	5	L	N	5
train	16774	5	N	S	5
hiver	8257	5	R	S	5
repas	5671	5	S	T	5
sport	1625	5	T	D	5
perdu	22952	5	U	S	5

APPENDIX D

List of the Stimuli Used in Experiment 4

Set 1 was used during the first two sessions (pilot experiment), and Set 2 during the following two sessions. Pseudowords of Set 2 were constructed from each test word by replacing a letter of the word with a letter that occurs at the same position in another test word (see bold letters in the Table). All but one pseudoword (AURSI) met this criterion.

	Critical Letter Position 1		Critical Letter Position 3		Critical Letter Position 5	
	Words	Pseudowords	Words	Pseudowords	Words	Pseudowords
Set 1	GENRE	TENRE	GENRE	GEDRE	GENRE	GENRU
	PETIT	BETIT	PETIT	PEFIT	PETIT	PETIR
	ODEUR	ADEUR	ODEUR	ODOUR	ODEUR	ODEUX
	ENVIE	ONVIE	ENVIE	ENSIE	ENVIE	ENVIL
	ROMAN	TOMAN	ROMAN	RODAN	ROMAN	ROMAS
Set 2	**ROMAN**	**S**OMAN	RO**M**AN	RO**J**AN	ROMA**N**	ROMA**T**
	AUSSI	**O**USSI	AU**S**SI	AU**R**SI	AUSS**I**	AUSS**E**
	SUJET	**A**UJET	SU**J**ET	SU**S**ET	SUJE**T**	SUJE**N**
	AUCUN	**R**UCUN	AU**C**UN	AU**D**UN	AUCU**N**	AUCU**T**
	ORDRE	**A**RDRE	OR**D**RE	OR**C**RE	ORDR**E**	ORDR**I**

APPENDIX E

Simulation of a General Visual Impairment

A right visual field deficit was simulated by decreasing the individual letter recognition probabilities given in Table 1 for letters placed to the right of fixation. Probabilities were decreased by a constant value of either .05, .10, .20, or .30. A higher value represents a stronger visual "impairment". The present table gives the theoretical probability of recognising a five-letter string with a right visual field deficit of .05. The resulting viewing position curve is presented in the left panel of Fig. 13.

Position of Fixated Letter in the String	Recognition Probabilities of Individual Letters of a Five-letter String					Probability of Recognising the Entire String
	1	2	3	4	5	
1	1.00	.90	.85	.80	.75	.46
2	.91	1.00	.90	.85	.80	.56
3	.82	.91	1.00	.90	.85	.64
4	.73	.82	.91	1.00	.90	.52
5	.64	.73	.82	.91	1.00	.35

APPENDIX F

Simulation of a Letter-processing Deficit

A deficit in processing letters of the second part of words was simulated by decreasing the recognition probabilities of the third, fourth, and fifth letter of the string, except when these letters were directly fixated or adjacent to the directly fixated letter. Probabilities were decreased by a constant value of either .20, .30, or .40. The simulation of a deficit in letter processing was based on the values given in Appendix E that resulted from the simulation of a right visual field deficit of .05. The following table gives the theoretical probability of recognising a five-letter string as a function of fixation position with a deficit in letter processing of .03. The resulting viewing position curve is presented in the right panel of Figure 13 circles.

Position of Fixated Letter in the String	\multicolumn{5}{c	}{Recognition Probabilities of Individual Letters of a Five-letter String}	Probability of Recognising the Entire String
	1	2	3	4	5	
1	1.00	.90	.55	.50	.45	.11
2	.91	1.00	.90	.55	.50	.23
3	.82	.91	1.00	.90	.55	.40
4	.73	.82	.91	1.00	.90	.49
5	.64	.73	.52	.91	1.00	.22

IMPLICIT VS. LETTER-BY-LETTER READING IN PURE ALEXIA: A TALE OF TWO SYSTEMS

Eleanor M. Saffran and H. Branch Coslett
Temple University, Philadelphia, USA

Recent studies of pure alexia present a contradictory picture. Despite evidence of impaired letter identification in letter-by-letter readers, some patients are able to carry out lexical decision and other tasks under conditions of rapid presentation, although they are seldom able to identify these stimuli explicitly. We review evidence for both facets of pure alexic performance and offer an account of this pattern in terms of right- and left-hemisphere reading mechanisms. Specifically, we suggest that the right hemisphere supports performance in covert reading tasks, and that letter-by-letter reading is the product of the left hemisphere, operating on information transmitted from the right.

INTRODUCTION

Reading can be severely impaired by damage to visual areas in the left hemisphere and/or to structures that provide input to the left-hemisphere visual system (e.g. Binder & Mohr, 1992; Damasio & Damasio, 1983). In the purest cases ("pure alexia", or "alexia without agraphia"), reading is seriously affected despite sparing of orthographic knowledge, demonstrated by the patient's ability to write and recognise words spelled orally. Most alexic patients achieve some measure of success in identifying printed words, albeit slowly and effortfully, by implementing a strategy of reading letter by letter (LBL). The hallmark of this approach is a monotonic relationship between the number of letters in a word and the time taken to read it.

Recent research has centred on two facets of the performance of LBL readers that appear contradictory. First, there is evidence for an impairment in letter identification that could account for the reading problem (e.g. Arguin & Bub, 1993; Behrmann & Shallice, 1995; Reuter-Lorenz & Brunn, 1990). At the same time, it has been shown that some alexic patients are able to access lexical information from briefly presented letter strings (e.g. Bub & Arguin, 1995; Coslett & Saffran, 1989a,

---

Requests for reprints should be addressed to Dr. Eleanor M. Saffran, Center for Cognitive Neuroscience, Department of Neurology, Temple University School of Medicine, 3401 North Broad Street, Philadelphia, PA 19140, USA (Tel: (215)707-3090; Fax: (215)707-3843; E-mail: saffran@vm.temple.edu).

We thank Marlene Behrmann, Max Coltheart and Marie Montant for helpful comments on the manuscript. Preparation of this paper was supported by grant number RO1 DC 00191-16 from the National Institutes of Health.

1989b; Howard, 1991; Shallice & Saffran, 1986). This capacity is demonstrated in lexical decision and semantic categorisation tasks (Coslett & Saffran, 1989a; Shallice & Saffran, 1986), and in the superior recognition of letters in words as compared with nonwords (the word superiority effect; e.g. Bub, Black, & Howell, 1989; Reuter-Lorenz & Brunn, 1990). As the patients are rarely able to report the stimuli presented in these tasks, their performance has been characterised as implicit (or tacit, or covert) reading[1]. To identify printed words explicitly, they resort to the cumbersome procedure of reading LBL, which requires longer exposure to the printed word, and may even interfere with implicit reading (Coslett & Saffran, 1994; Coslett, Saffran, Greenbaum, & Schwartz, 1993).

The co-occurrence of impaired letter identification and rapid but implicit processing of printed words presents a paradox. If patients are indeed impaired in letter identification, how are they able to process words rapidly? And why is it that this capacity to process words rapidly does not lead to explicit report?

Several years ago, we put forward an account that attempts to resolve these contradictory findings (Coslett & Saffran, 1989a, 1994). Our proposal is similar, in some respects, to that of Déjèrine (1892; Bub, Arguin, & Lecours, 1993), who first described the syndrome of pure alexia. According to Déjèrine, the reading deficit is the result of the isolation of left-hemisphere language mechanisms from visual input. This condition arises from damage to posterior portions of the left hemisphere that disrupts the transmission of visual information to areas that mediate word recognition. The transmission problem applies to right- as well as left-hemisphere visual input. As left-hemisphere visual areas are generally affected, early stages of visual processing are in most cases restricted to the right hemisphere. Our account incorporates this assumption. We suggest, further, that LBL reading is a compensatory strategy adopted by the left hemisphere, which attempts to recognise printed words despite its inability to access orthographic information directly. The left hemisphere reads via a slow, serial process of letter identification, based on information transmitted by the right hemisphere. Inefficient though it may be, this procedure supports explicit identification and pronunciation of printed words through activation of the left-hemisphere lexical system. The covert reading evidenced under rapid presentation conditions is assumed to be the product of a different reading system. This system, based in the right hemisphere, does not articulate directly with left-hemisphere mechanisms for language production. Hence it does not support explicit identification of the word.

We will return to this account later on. First, we examine the set of phenomena that needs to be accommodated by a model of reading in pure alexia. We begin with a discussion of the evidence for impaired letter identification, and then turn to the demonstrations of covert reading.

---

[1] See Feinberg, Dyches-Berke, Miner, and Roane (1994) for a different view of this behaviour.

LETTER PROCESSING IN PURE ALEXIA

Putting aside, for the present, the complications introduced by evidence for tacit reading, there are good reasons to link the reading deficit in pure alexia to an impairment in letter identification. Some alexic patients, including the subject of Déjèrine's (1892; Bub et al., 1993) original report (Monsieur C), are completely unable to identify letters on the basis of visual input (e.g. Binder & Mohr, 1992; Coslett & Saffran, 1989a, 1992). More commonly, alexics retain some capacity to name letters, which they can employ to identify words letter by letter. In many cases, the LBL strategy is handicapped by letter naming errors, which often reflect a visual similarity to the target (e.g. Patterson & Kay, 1982).

The letter-processing deficit in pure alexics has generated a good deal of interest. Among the issues that have been addressed are the following: (1) Are the patients impaired in recognising single letters, or does the problem arise only with letter arrays? (2) Does the deficit apply only to letters, or does it reflect a more general visual processing impairment?

Evidence for the deficient processing of letter arrays dates back to the 1960s. Kinsbourne and Warrington (1962) found that alexic patients demonstrated normal tachistoscopic recognition thresholds for single letters, but were severely impaired in identifying strings of letters. Performance was affected only by the number of stimuli, and not by their size, position, or spatial separation. Successive presentation of the letters eliminated the problem, but only when the interstimulus interval was increased substantially. Although the patients were similarly impaired in identifying arrays consisting of silhouettes of objects, they performed normally on a dot-counting task carried out under tachistoscopic conditions. Kinsbourne and Warrington concluded that the deficit did not involve simultaneous perception per se, but rather the parallel processing of complex visual forms. Like Kinsbourne and Warrington (1962), Levine and Calvanio (1978) found normal recognition thresholds for single letters in pure alexics, as well as normal performance on a backward masking task with single letters. In contrast, the patients were grossly deficient on letter arrays. These authors suggested that the deficit reflects malfunctioning of a buffer that maintains visual stimuli until they are recognised. Farah and Wallace (1991) argued for a general visual deficit, based on their finding that the speed of letter-by-letter reading by a pure alexic interacted with stimulus quality, manipulated by presenting words with a superimposed line pattern. Friedman and Alexander (1984) offered a similar argument, supported by evidence that pure alexics' thresholds for recognising line drawings were elevated relative to those of normal subjects. Although some recent studies have provided evidence of a general visual processing impairment in LBL readers (e.g. Sekuler & Behrmann, 1996), this finding has not been obtained in others (e.g. Arguin & Bub, 1993)[2].

---

[2] One possible explanation for the visual deficits observed in these patients may be the loss of input from left-hemisphere visual areas. Bilateral impairments due to unilateral occipital lesions have been demonstrated, possibly reflecting a decrease in visual attention (Rizzo & Robin, 1996).

Irrespective of the origin of the deficit—whether it is specific to letters or not—it is clear that letter identification poses a significant problem for these patients. Compelling evidence comes from the cross-case letter-matching task, which requires subjects to determine whether two letters (e.g. Rr) have the same name. To perform this task, letter information must be processed to a level at which form is no longer relevant. Although pure alexics perform well on same-case identity trials (e.g. RR), they are often deficient on cross-case identity (Rr) matching (Bub et al., 1989; Kay & Hanley, 1991; Reuter-Lorenz & Brunn, 1990). Other evidence for an impairment in letter identification comes from a study by Behrmann and Shallice (1995), whose subject demonstrated slow processing not only with letter arrays but also (as in the Kinsbourne and Warrington [1962] study) when single letters were presented sequentially. On the basis of such findings, a number of investigators have concluded that the problem for LBL readers arises in the conversion of letter shapes to abstract letter representations (e.g. Arguin & Bub, 1993; Behrmann & Shallice, 1995; Reuter-Lorenz & Brunn, 1990).

The problem in letter identification could well account for the slow pace of LBL reading, and for the effect of word length on oral reading. But this impairment is difficult to reconcile with the evidence summarised in the next section, which demonstrates that some alexic patients are capable of processing printed words rapidly, although they are generally unable to report them.

EVIDENCE FOR IMPLICIT READING IN PURE ALEXIA

In the 1970s, investigators began to report instances of spared reading in pure alexics. For example, Caplan and Hedley-White (1974) described a patient who was impaired at letter naming and had great difficulty reading words aloud. When shown a letter string that contained a word plus extraneous letters, the patient was able to indicate which letters were not part of the word. She also matched auditory to written words at above-chance levels and succeeded in pointing to words that belonged to a designated semantic category. Albert, Yamadori, Gardner, and Howes (1973) reported similar phenomena, in a patient who was also able to match printed words to objects. Landis, Regard, and Serrat (1980) described another pure alexic who succeeded in matching tachistoscopically presented words to objects, but appeared to lose this ability as he began to read letter by letter. Additional examples were reported by Grossi, Fragassi, Orsini, De Falco, and Sepe (1984) and Kreindler and Ionescu (1961).

The first extensive report of covert reading in pure alexia was provided by Shallice and Saffran (1986). Their patient, ML, was able to carry out lexical decision and semantic categorisation on words that were presented too briefly (2 seconds)[3] to support LBL reading.

---

[3] The 2-second rate was used because the stimuli were presented on cards and it was considered that faster rates could not be timed adequately.

Notable features of ML's performance include the following: above-chance but far from perfect performance on lexical decision, with error rates dependent on word frequency and the similarity of nonwords to words; and performance ranging from 70 to 94% correct on binary choice categorisation tasks, where the categories included countries (in vs. out of Europe), occupations (author vs. politician), and objects (living vs. nonliving). ML was also able to locate places, corresponding to names presented for 2 seconds, on an outline map of Britain. In contrast, he was rarely able to report the letter string under these presentation conditions, although he occasionally volunteered some information about the word (e.g. for *knife*, "tool or something like that... hardware of some kind... don't know what its function is... but know would use it with the hand").

Another notable feature of ML's tacit reading behaviour was a lack of sensitivity to the appropriateness of affixes. In contrast to his relatively good lexical decision performance on morphologically simple words, he performed poorly with inflected words, accepting stimuli with inappropriate affixes (e.g. *windowing, strongs, galloply*) as often as stimuli with appropriate affixes (e.g. *windows, strongly, galloping*). ML's performance with affixed stimuli was not the result of failure to process the ends of words; he proved to be quite sensitive to nonwords consisting of words plus endings that are not legal affixes although they occur quite frequently in English (e.g. *pockete, tightent, strangey*).

A few years later, we carried out a similar set of studies on four LBL readers (Coslett & Saffran, 1989a). Under conditions of brief presentation, these patients proved to be sensitive to words on lexical decision tasks, and to their meanings on tests of semantic categorisation and word–picture matching. We also obtained evidence of tacit reading in three additional cases. One (JWC) was another LBL reader (Coslett et al., 1993). The other two patients (CB and EM) were unable to name letters at all, and had difficulty identifying other stimuli to visual presentation; as they could name to definition and to tactile presentation, they were classified as optic aphasics (Coslett & Saffran, 1989b, 1992).

The performance of these seven patients on lexical decision and binary-choice categorisation tasks is summarised in Tables 1 and 2. All of them performed at above chance levels on both tasks, although their scores generally fell below the levels expected of normal subjects. Six of the patients were also tested on a word-to-picture matching test; those data can be found in Table 3. Again, they achieved a fair degree of success. As was the case for Shallice and Saffran's (1986) ML, our subjects reported a negligible proportion of these letter strings, although they occasionally volunteered information that indicated access to their meanings. Words were most often named on the categorisation tasks, but these responses were wrong at least as often as they were correct.

Other tasks administered to this set of patients included the affixed lexical decision task performed by ML (Shallice & Saffran, 1986). The data are summarised in Table 4, where it is evident that their performance patterns were similar to ML's. All six subjects reliably dis-

Table 1. Lexical Decision at Brief Exposures (Reprinted from Coslett & Saffran, 1994)

| | | % "Yes" Responses |||||| |
| | | Words ||| Nonwords ||| |
Patient	Exposure (Msec)	HiFreq N = 60	LoFreq N = 60	Hi + Lo N = 120	Hi – N N = 60	Λo – N N = 60	Hi + Lo N = 120	d'
Letter-by-letter								
JG	250	75	37	56	35	25	30	0.68*
TL	150	82	78	80	63	38	50	0.84*
JC	250	78	52	65	28	11	19	1.30*
AF	250	80	53	67	42	18	30	0.96*
JWC	249	93	48	71	57	43	45	0.55*
No letter ident								
EM	250	92	58	75	20	12	16	1.63*
CB	unlim	88			34			1.59*

Note: Hi–N refers to word-like nonwords, Lo-N to nonwords with unusual orthographic patterns.
*P < .05.

Table 2. Category Decision at Brief Exposures (Reprinted from Coslett & Saffran, 1994)

| | | % Correct |||||
| | | Animal? |||| Edible? |
Patient	Exposure (msec)	Animal (e.g. Mouse) N = 25	Visual Foil (e.g. Mount) N = 25	Unrel. Foil (e.g. Rally) N = 25	Total N = 75	Total N = 75
Letter-by-letter						
JG	250	80	72	72	75	80
TL	100	96	52	60	69	67
JC	250	68	92	96	85	79
AF	250	80	60	72	71	76
JWC	249	80	68	72	73	76
No letter ident						
EM	250	100	92	100	97	96
CB	unlim.	84	76	84	81	79

Table 3. Word-to-picture Matching at Brief Exposures (Reprinted from Coslett & Saffran, 1994)

		Per cent Correct	
Patient	Exposure (msec)	Different Category N = 62 (house/horse)	Within Category N = 32 (snake/snail)
Letter-by-letter			
JG	500	87	90
TL	100	69	...
JC	250	74	87
AF	250	66	73
JWC	249	84	...
No letter ident			
EM	250	87	88

criminated unsuffixed words from unsuffixed nonwords ($P < .05$); only one (EM) performed significantly better on appropriately than inappropriately suffixed words, but even he accepted two thirds of the latter. EM was also given rhyme detection tasks that examined his sensitivity to the phonological properties of printed words (Coslett & Saffran, 1992). In one task, two words were presented successively on a computer screen; in the other, one word was presented auditorily and the second visually. EM performed at chance on the first task and failed to complete the second one, claiming that it "simply can't be done".

Additional evidence of the capacity to process letter strings in parallel comes from demonstrations of the word superiority effect (WSE) in LBL readers. The WSE refers to the superior recognition of letters in words as compared with strings of unrelated letters. Several investigators have demonstrated WSEs in LBL readers tested under brief presentation conditions (Bowers, Bub, & Arguin, 1996; Bub et al., 1989; Friedman & Hadley, 1992; Reuter-Lorenz & Brunn, 1990). It is noteworthy that the patients display different patterns of accuracy

Table 4. Sensitivity to the Appropriateness of Affixes (Reprinted from Coslett & Saffran, 1994)

			% "Yes" Responses				
			Lexical Root Morphemes			Nonlexical Roots	
Patient	Exposure (msec)	N	No Suffix (broad)	Correct Suffix (broadest)	Incorrect Suffix (broading)	No Suffix (narrid)	Suffixed (narridest)
Letter-by-letter							
JG	250	63	63	63	65	19	37
TL	100	40	88	78	75	13	13
JC	250	63	84	79	73	40	32
AF	250	40	85	75	65	18	18
No letter ident							
EM	250	63	83	87	67	10	10
CB	unlim.	40	93	63	68	30	20

across letter positions than they do when reading LBL. Bub et al.'s (1989) patient showed a flat pattern with words, whereas the patient reported by Reuter-Lorenz and Brunn (1990) demonstrated the normal U-shaped pattern, reflecting better performance on the first and last letters of the string. These patterns are in marked contrast to the monotonic decrease in accuracy across letter positions that is characteristic of LBL reading (e.g. Friedman & Hadley, 1992).

Some investigators who have tested alexic patients at short exposures have failed to obtain evidence for covert reading (e.g. Behrmann & Shallice, 1995; Patterson & Kay, 1982)[4]. Such failures are not surprising, as it is not easy to gain patients' cooperation in these tasks. Some of our subjects have claimed that they could not see the letter string under tachistoscopic conditions. They also found it difficult to abandon the LBL strategy that allowed them to identify words explicitly. When letter strings were presented only briefly, they sometimes attempted to read letter by letter and succeeded in reporting only one or two letters. We found it necessary to actively discourage the patients from engaging in this type of reading and to urge them, instead, to try to gain an appreciation of the letter string as a whole. In the course of performing the task, we repeatedly admonished them not to try to report the letters. We also encouraged them to venture a guess if they were uncertain. In contrast, other investigators have stressed report of the letter string as the principal task, and have instructed their subjects to make lexical decision or categorisation judgements only when they are unable to identify the word (e.g. Patterson & Kay, 1982). The patient's experience in the testing situation may also be a critical factor. If patients undergo extensive testing on tasks that require them to report the stimuli, they may have difficulty abandoning the LBL strategy even when it cannot be used effectively. In our own work with alexic patients, we were quick to initiate testing under brief presentation conditions that are incompatible with LBL reading.

ML's behaviour suggested that the LBL strategy might even conflict with the faster mode of word identification that supported his performance on lexical decision and categorisation tasks (Shallice & Saffran, 1986; also see Landis et al., 1980). Although stimuli were exposed for 2 seconds, ML rarely took advantage of the full presentation time when performing these tasks; instead, he would glance briefly at the letter string and look away. In contrast, he took all the time allowed in explicit report tasks, where he showed the effects of word length associated with LBL reading.

To test the hypothesis that implementation of the LBL strategy interferes with covert reading, we carried out additional studies with patient JWC (Coslett & Saffran, 1994; Coslett et al., 1993). We examined JWC's ability to cate-

---

[4] Behrmann and Shallice's patient, a relatively fast letter-by-letter reader, was asked to perform a semantic categorisation task of a 500msec exposure. She read over half of the words correctly, and performed at chance in categorising the remainder. It is conceivable that this presentation rate was too slow to discourage letter-by-letter reading.

gorise words (edible or not) under two different conditions: short exposure (250msec) and longer exposure (2000msec). In the latter condition, he was asked to identify words explicitly, and to categorise the stimuli he could not name. The results are summarised in Table 5. JWC performed reasonably well on judgement tasks at short exposures, although he was seldom able to identify the stimuli explicitly. In contrast, when performing under explicit report instructions, JWC was unable to categorise words he failed to name.

Other evidence that suggests that explicit and implicit reading are based on different procedures comes from an analysis of the effect of word length on lexical decision judgements. We examined this relationship in four patients who showed an effect of word length on reading time under free presentation conditions (that is, they all met the criteria for letter-by-letter reading; see Coslett & Saffran, 1994). In contrast to their performance when reading letter-by-letter, the length of the letter string had no effect on accuracy in the lexical decision task performed under tachistoscopic conditions. Evidence of a similar nature has been reported by Bub and Arguin (1995). Other relevant data come from an examination of the effect of exposure duration on the lexical decision performance of patients AF and JG (Coslett & Saffran, 1989a). Whereas AF was 75% correct at 250msec, she performed at chance when stimuli were presented for 2000msec. JG's performance across a range of exposures is summarised in Table 6. His performance declined to chance levels at 2000msec, an exposure that appeared to be long enough to apply the LBL strategy but too short to use it successfully.

These results indicate that the procedures pure alexics employ for the purpose of identifying letter strings explicitly do not support—and may even interfere with—their ability to perform lexical decision and catego-

Table 5. Category Decision and Naming Performance of JWC at Long and Short Exposures (Reprinted from Coslett & Saffran, 1994)

Task Order	Task	Exposure	N	No. Named	% Correct
1	Categorisation male or female name?	249 msec	30	1–	100
2	Categorisation animal?	249 msec	75	3 +	85
3	Name word	Unlimited			(only trials not named)
4	Name, then edible?	2000 msec	25 foods	16 +, 3–	
			25 vis. foils	0 +, 12–	
			25 unrel. foils	0 +, 4–	
			Total		51
5	Categorisation	249 msec	25 foods	3 +	
			25 vis. foils		
			25 unrel. foils	1 +	
			Total		86

Table 6. Lexical Decision in JG (Percent Yes Responses) as a Function of Exposure Time (Reprinted from Coslett & Saffran, 1994)

	\multicolumn{5}{c}{Exposure Time in msec}
	50	250	500	2000	Unlimited
Words	49%	59%	55%	51%	86%
Hi F	57	68	70	64	
Lo F	40	57	40	37	
Nonwords	22%	36%	21%	54%	8%
Hi N	27	33	25	68	
Lo N	18	28	18	40	
$d' =$	0.75	0.72	0.93	0.13	2.49
$z =$	4.36	4.40	5.38	0.775	7.32

risation judgements. Although some investigators have failed to demonstrate this covert reading capacity, there is evidence that it occurs with some frequency in LBL readers. How, then, do we account for the conjunction of impaired letter identification and implicit reading?

EXPLICIT AND IMPLICIT READING: TWO SYSTEMS

One possible account of this pattern is that patients apply different response criteria in explicit and implicit tasks: Although not confident about the identity of a briefly exposed word, they are nevertheless willing to make judgements of lexicality or category membership. While it is difficult to exclude this possibility, the evidence of incompatibility between the two types of responses argues against it. Under conditions where a sizeable proportion of words can be identified explicitly, there should be a significant number of stimuli that do not meet the subject's criterion for explicit identification although they have been processed sufficiently to satisfy criteria for lexicality or category membership. Hence it should be possible to make judgements about stimuli that cannot be identified with confidence. To the contrary, JWC performed at chance on semantic categorisation at the same time that he was reading 51% of the words correctly. Behrmann and Shallice's (1995) patient DS showed the same pattern, failing to categorise words that were not explicitly identified while reading 55% of the words correctly.

As noted in the Introduction, our account of the co-occurrence of LBL and implicit reading posits two separate mechanisms, one based in the left hemisphere that allows explicit word identification, and another in the right hemisphere that supports covert reading. The rationale for the two-system hypothesis is set out below.

It is a reasonable supposition that the damage that gives rise to pure alexia impedes the transmission of visual information to left-hemisphere areas that subserve word recognition[5]. Although there are some cases in which

---

[5] Impairment to the visual word form system is an alternative view of letter-by-letter reading (Warrington & Shallice, 1980). The presence of surface dyslexia (a strong orthographic regularity effect) in letter-by-letter reading would suggest such an impairment. Although some letter-by-letter readers do manifest such symptoms (e.g. Bowers, Bub, & Arguin, 1996; Patterson & Kay, 1982), most do not.

the lesion does not impinge on occipital cortex per se (see Black & Behrmann, 1994, for evidence on this point), pure alexia is generally accompanied by a right visual field deficit, in some instances partial and in others complete (e.g. Damasio & Damasio, 1983). The visual defect results from damage to the left occipital cortex, and/or to structures that provide input to left-hemisphere mechanisms; these include the lateral geniculate, optic radiations, and callosal fibres that project to the left hemisphere from visual areas in the right hemisphere (Binder & Mohr, 1992; Damasio & Damasio, 1983). One consequence of this damage is that the visual information that supports LBL reading is processed—up to a point, at least—by the right hemisphere. This condition is likely to hold, as well, for the rare patients who do not have right visual field defects. Since naming a printed word involves capacities that are typically localised in the left hemisphere, explicit identification of a written stimulus entails the transfer of information from the right to the left hemisphere. LBL reading reflects the nature of this information, which has yet to be characterised.

The variety of performance patterns on written words and other visual stimuli observed in patients with right visual field defects—particularly those with right homonymous hemianopias—is relevant to this issue. Some of these patients do not become alexic, presumably because left-hemisphere orthographic mechanisms remain accessible to visual input. At the other extreme are patients who are unable to name letters as well as other types of visual stimuli—those who are optic aphasic as well as pure alexic; in such cases, left-hemisphere language mechanisms would appear to have little or no contact with visual input. Still others are alexic but retain the ability to name letters as well as other visual forms. As these patterns presumably reflect differences in lesion site, a consideration of these differences should be informative. Unfortunately, as Black and Behrmann (1994) have observed, investigators who have focused on the anatomic basis of pure alexia have rarely obtained adequate behavioural data, and those with a cognitive orientation have not been much concerned with anatomy.

Some evidence that bears on this issue has been assembled by Binder and Mohr (1992), who limited their purview to the processing of letters and words. These authors examined the lesion sites of 17 patients, with left posterior cerebral artery infarctions, who were divided into 3 groups: those with normal reading ability; "global" alexics, who were unable to read at all; and "spelling" alexics, or LBL readers. The patients with preserved reading incurred damage to medial and ventral areas of the occipital lobe, sparing dorsal white matter pathways as well as ventral temporal areas implicated in the visual processing of objects and words (e.g. Beauregard et al., 1997). The patients with global deficits had larger lesions that included callosal pathways (splenium and forceps major) as well as dorsal white matter above the lateral ventricles and posterior to the forceps major. In these cases, it is reasonable to assume that the lesion blocked the transmission of visual information from the right to the left hemisphere, including letter information.

The patients identified as LBL readers had lesions that included ventral temporal lobe areas concerned with later stages of visual processing, but sparing the dorsal regions implicated in global alexia. The preservation of dorsal white matter may underlie their ability to transmit letter information from the right to left hemisphere.

Data from our own series are not entirely consistent with these findings. The lesion sites in our LBL readers varied considerably; only two of the five patients had sustained damage to the ventral temporal region involved in the cases of spelling alexia identified by Binder and Mohr. This variability is not surprising, as the functional deficit could arise from any lesion, or combination of lesions, that prevents visual input from accessing left-hemisphere orthographic processing mechanisms[6]. For example, one of our cases (JG) became alexic as a result of two lesions, one to the left lateral geniculate, which largely deafferented left-hemisphere visual cortex, and a callosal lesion that presumably restricted visual input from the right hemisphere. There is one respect, however, in which our data are quite compatible with the findings of Binder and Mohr. The two patients in our series who showed no capacity to identify letters (CB and EM) incurred damage to the dorsal white matter implicated in Binder and Mohr's global alexics. Figure 1 displays CT images for EM and for JWC, the LBL reader with the lesion pattern most similar to that of EM. It is evident that the lesion spared the dorsal white matter in JWC, as it did in our other LBL readers. One inference that might be drawn from these observations is that the ability to read LBL depends on the preservation of dorsal pathways that allow the transmission of some visual information from the right to the left hemisphere.

In developing our account of the performance pattern in pure alexia, we have considered not only the behaviour of LBL readers but also that of two of our patients (CB and EM) who were completely unable to name letters and had difficulty naming other visual stimuli as well. CB (Coslett & Saffran, 1989b) was a severe optic aphasic who was virtually unable to name objects to visual presentation (the single exception was a relatively obscure item [yoke] on the 80-item Boston Naming Test). The majority of CB's responses in confrontation naming tasks were confabulatory and perseverative in nature. EM (Coslett & Saffran, 1992) was somewhat less impaired in naming objects. Unlike CB, he provided semantic information for objects he could not name and occasionally some semantic information about printed words that he could not explicitly

---

[6] As noted by Damasio and Damasio (1983) and Bub et al. (1993), Déjèrine initially attributed the failure to transfer graphemic information from the right to the left hemisphere to disruption of the white matter tracts in the left hemisphere (forceps major). Although lesions of the corpus callosum can disconnect the visual association cortices of the two hemispheres, the same fibres can be disrupted following lesions within the substance of the hemisphere, which is frequently the site of the lesion in pure alexia. Differences in the location and size of these lesions must account, in part, for the variability in this patient population.

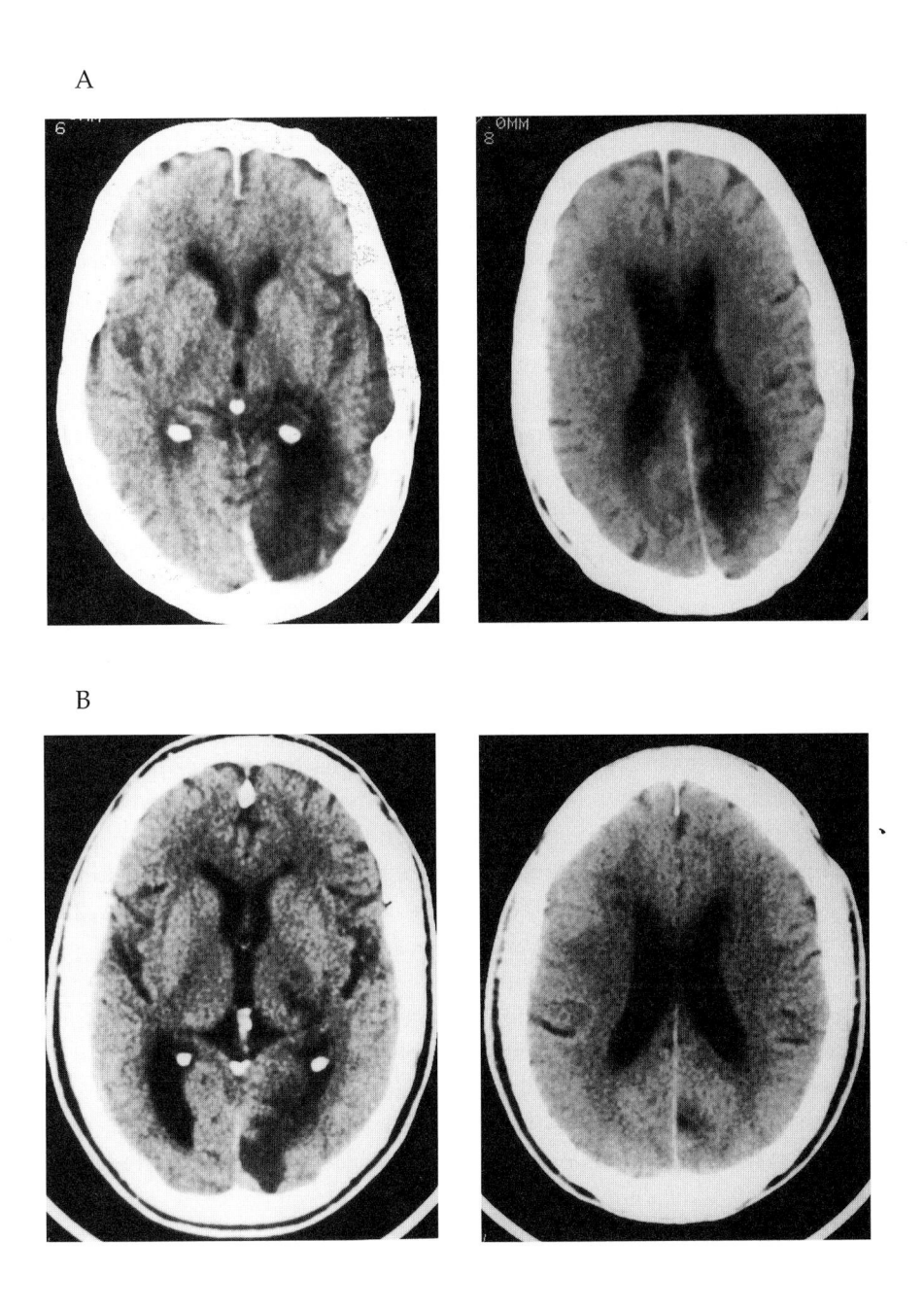

Fig. 1. CT scans from patients EM (A) and WC (B). Note the involvement of dorsal white matter in EM's lesion but not in WC's.

identify (e.g. *cookies* → "candy, a cake"; *bacon* → "a bad thing, bad for your body"). In these cases, we assumed that visual information processing was restricted to the right hemisphere, and that callosal transfer was limited to semantic information, a route that was evidently more effective for EM than CB. Letters, which have no conceptual content, could not be named. Because letter information could not be transferred, these patients were unable to identify words explicitly. It is important to note, however, that they demonstrated the same covert reading patterns as the five patients who were capable of reading letter-by-letter.

Whereas some LBL readers exhibit symptoms of optic aphasia (e.g. Shallice & Saffran, 1986), visual confrontation naming is reported as relatively unimpaired in most of the published cases, at least as measured by the content of their responses. Some patients have responded with long latencies, particularly to complex visual forms (Sekuler & Behrmann, 1996), or with difficulty when multiple objects were presented (Kinsbourne & Warrington, 1962). Although object naming might be sustained by the transfer of semantic information from the right hemisphere, this mechanism would not support the naming of letters. The ability to name letters as well as objects suggests that some visual information is passing from the right to the left hemisphere. Furthermore, as noted earlier, the letter identification errors of LBL readers appear to be visually based (e.g. Patterson & Kay, 1982), reflecting either some degradation in the transfer of visual information or impairment in the letter naming process per se. At present, we can only speculate on the nature of this letter identification process. One possibility is that letters are recognised by the same mechanisms that are used to identify visual objects (Coslett & Saffran, 1994). Compared with normal modes of orthographic processing, such a procedure is likely to be serial rather than parallel, and therefore relatively slow.

We are assuming that the information that the left hemisphere uses to support LBL reading is visual in nature. Other alternatives seem less likely. Letters do not have semantic content, so the information transferred from the right hemisphere is unlikely to be semantic in nature. We have argued elsewhere that direct linkages between letter representations in the two hemispheres are also unlikely, on the following grounds: The vast majority of callosal connections appear to be homotopic (e.g. Pandya & Seltzer, 1986). In postulating links between information-processing modules in the two hemispheres, it is therefore reasonable (Coslett & Saffran, 1994, p. 326) to limit connections to modules of the same general type, adopting "a conservative stance with respect to the types of functional systems that are likely to be homotopically represented". This entails assuming "homotopic representation only for functional systems that have a venerable phylogenetic history". On these grounds, we should not "assume homologous representation for the mechanisms that subserve written language, which is phylogenetically recent, culturally determined, ... generally acquired with some effort—and hence likely to be variable in its anatomic substrate".

We have attributed the implicit reading performance of pure alexics to mechanisms that reside in the right hemisphere[7]. A critical piece of the evidence in support of the right-hemisphere account is the very similar performance of our LBL and non-LBL readers (CB and EM) on covert reading tasks. Anatomical as well as behavioural evidence indicates that interhemispheric transmission pathways were significantly damaged in these two patients, who were unable to name visually presented objects or letters. Hence there are grounds for assuming that visual information processing was limited to the right hemisphere in these cases. If the right hemisphere supports the covert reading performance of CB and EM, it is reasonable to assume that it also supports the covert reading performance of alexic patients who can read letter by letter.

Figure 2 illustrates the bi-hemispheric model that we have just outlined. Tacit recognition, probed in lexical decision and categorisation tasks, is subserved by the right hemisphere, whereas LBL reading depends on left-hemisphere mechanisms that operate on processed visual information transmitted from the right hemisphere. In an earlier paper (Coslett & Saffran, 1994), we pointed out that tasks that require an explicit verbal response are likely to mediated by the left hemisphere, which is likely to take control of the task in such circumstances. To perform tacit reading tasks, some consequences of right-hemisphere lexical processing must be available to conscious awareness. To access this information, it is presumably necessary to direct attention to the right hemisphere, which may be difficult if the left hemisphere is attempting to name the word. We offer this as one possible reason for the failure to demonstrate tacit reading performance in some LBL readers. Data from a patient with a lesion to the posterior portion of the corpus callosum are relevant to this point (Faure & Blanc-Garin, 1994). The patient's ability to categorise stimuli presented to the left visual field improved significantly when the semantic task was preceded by a visuospatial task, which presumably activated the right hemisphere. Performance on left visual field stimuli also improved under dual-task conditions; the requirement to perform a concurrent verbal memory task was presumed to "overload" the left hemisphere, allowing the right hemisphere to assume control of categorisation performance.

---

[7] As one reviewer has pointed out, there is evidence that the right hemisphere processes letter information serially and not in parallel. Young and Ellis (1985; Ellis, Young, & Anderson, 1988) demonstrated that word length affected oral reading, lexical decision, and semantic categorisation of left visual field stimuli in normal subjects, and Reuter-Lorenz and Baynes (1992) found that word length affected lexical decision as well as the word superiority effect in the right hemisphere of a callosotomy patient. Ellis et al. attribute the word length effect in normals to the transfer of letter information to the left hemisphere; this would not, however, account for these effects in the performance of the isolated right hemisphere. There is no evidence of word length effects in the implicit reading we have attributed to the right hemisphere, although it should be noted that these studies have generally been carried out with fairly long exposure durations. The impact of word length on covert reading clearly requires further investigation.

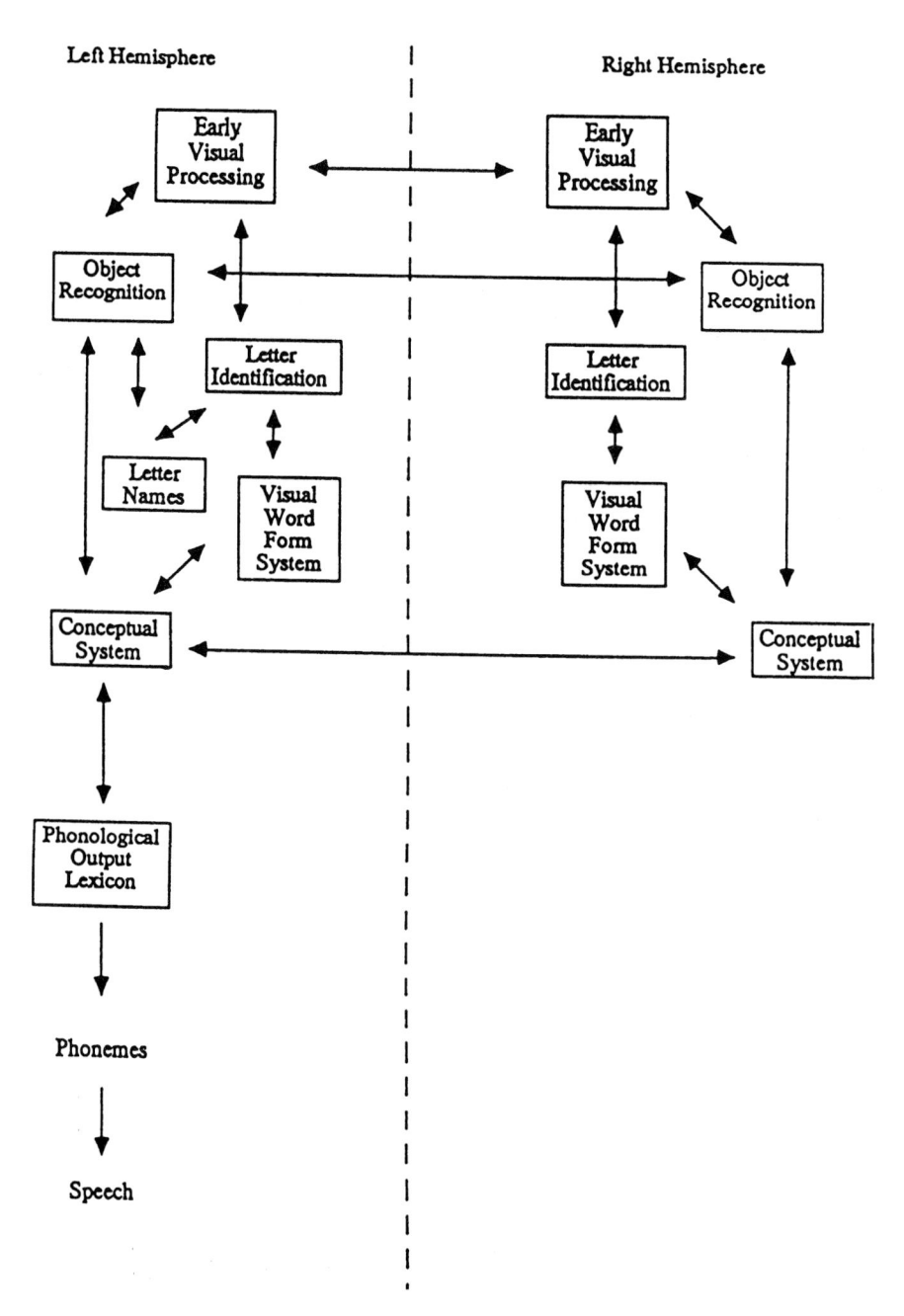

Fig. 2. An information processing account of the hemispheric processes involved in single word reading (Reprinted from Coslett & Saffran, 1994).

Our account of reading in LBL patients is very similar to Cohen and Dehaene's (1995) interpretation of the performance of alexic patients on number processing tasks. Difficulty with numbers in pure alexia was noted by Déjèrine (1892), who observed that Monsieur C was able to report single digits but erred on multiple-digit numbers. The same pattern held for the two patients studied by Cohen and Dehaene. But whereas these patients failed to identify two-digit numbers correctly, they had no difficulty at all judging the relative magnitude of pairs of these numbers. To account for this performance pattern, Cohen and Dehaene hypothesise that magnitude comparisons are carried out by the right hemisphere, whereas explicit report is based on information transferred to the left hemisphere. Also worth noting is their patients' strong tendency to perseverate in tasks that required the report of two-digit numbers. Perseverative responses are also prevalent in optic aphasics in the context of confrontation naming tasks (e.g. Coslett & Saffran, 1989b; Lhermitte & Beauvois, 1973). Cohen and Dehaene cite Geschwind's (1965, p. 590) interpretation of this behaviour as "attempts to fill gaps in the information available to [the] speech area" when it is deprived of adequate input.

In further support of the hypothesis that the right hemisphere mediates covert reading in pure alexics, we note that their performance on these tasks parallels that of the isolated right hemisphere in commissurotomy and left-hemispherectomy cases. Although this body of evidence is subject to caveats having to do with the history of brain abnormalities in such patients, which might result in nonstandard lateralisation patterns, the parallels are nevertheless striking. Where right-hemisphere language capacity has been demonstrated in split-brain and left-hemispherectomy cases (see Baynes, 1990, for review), there is evidence that the right-hemisphere lexicon is biased toward high-frequency, concrete words and that the phonological and syntactic capacities of the right hemisphere are limited (e.g. Patterson, Vargha-Khadem, & Polkey, 1989; Zaidel & Peters, 1981). For example, the reading performance of the best-studied left-hemispherectomy case showed a bias toward concrete words and an inability to encode print phonologically; the patient also generated semantic errors in reading words (Patterson et al., 1989). Another relevant observation was recently reported by Michel, Henaff, and Intrilligator (1996) in a patient with damage to the posterior portion of the corpus callosum that spared both occipital lobes. This patient was able to read normally when stimuli were presented to the right visual field. With left visual field presentation, he had great difficulty reading nonwords, showed effects of word class and imageability in reporting words, and generated a substantial number of semantic errors. In contrast, the patient performed well on left visual field stimuli in lexical decision and categorisation tasks, although latencies were longer than with right visual field presentation. When asked to decide if two words rhymed, the patient achieved a high degree of success with right visual field stimuli, but performed randomly to stimuli in the left visual field.[8]

Of particular relevance to the right-hemisphere account is the alexics' apparent insensitivity to the appropriateness of inflectional affixes under conditions of rapid presentation (Table 4). This finding is consistent with evidence that syntactic abilities are relatively undeveloped in the right hemisphere (see Baynes, 1990, for review). Other pertinent data come from LBL readers who showed some recovery of explicit word identification under tachistoscopic conditions. The performance of three such patients is documented in Coslett and Saffran (1989a). Over the course of weeks to months of testing on covert reading tasks, these patients showed an increasing ability to report words. However, their performance was marked by biases that were not evident under unlimited exposure, which allowed them to read letter by letter. The patients performed significantly better on nouns than function words, on high- vs. low-imageability words, and on "pseudoaffixed" words (e.g. *flower*) vs. frequency-matched affixed words (e.g. *flowed*). In reading the affixed set, inflectional affixes were often omitted. Similar effects of word class were observed in a fourth patient, JWC, when he performed a lexical decision task under tachistoscopic conditions (Coslett et al., 1993).

Also of interest is Buxbaum and Coslett's (1996) recent report of a patient (JH) who showed characteristics of both letter-by-letter reading and deep dyslexia. From the outset, JH showed rapid reading of some words and slow, letter-by-letter reading of others, a process that was impeded by letter-naming difficulty. When reading words presented at brief exposures, JH produced a fair number (12%) of semantic errors. In reading aloud, he also showed part of speech and imageability effects, as well as difficulty with affixes and nonwords. Perhaps the most interesting finding was an effect of word length on the accuracy as well as latency of word reading; this effect interacted with imageability, such that the length effect was substantially greater with low-imageability words. This interaction was not observed when words were spelled aloud to the patient. On the basis of these data, it appears that JH used two quite different procedures for reporting printed word, one of which was effective only for high-imageability, non-affixed words. One possible interpretation of JH's performance is that the imageability and part-of-speech effects reflect processing carried out by the right hemisphere; these products of right-hemisphere processing, transmitted in the form of semantic information, might then serve to facilitate LBL-based word recognition in the left hemisphere. This is also a possible explanation for the word class effects noted earlier, which were observed in some of the LBL readers studied by Coslett and Saffran (1989a). These effects de-

---

[8] The pattern of left visual field/right hemisphere performance is similar to the reading performance of deep dyslexics, which, according to some accounts, is mediated by the right hemisphere (e.g. Coltheart, 1980; Cossu, da Prati, & Marshall, 1995; Saffran, Bogyo, Schwartz, & Marin, 1980; Schweiger, Zaidel, Field, & Dobkin, 1989).

veloped over a period of months, possibly reflecting an increased ability to transfer and/or utilise semantic information generated by the right hemisphere.

The most direct test of right hemisphere involvement in the reading performance of pure alexics was carried out by Coslett and Monsul (1994). They examined the effect of transcranial magnetic stimulation on the reading performance of JG, one of the patients studied by Coslett and Saffran (1989a, 1994). Transcranial magnetic stimulation, which induces a transient electrical current in underlying tissue, was applied to the right or left temporo-parietal area (spared in JG) while he was attempting to read words presented for 100msec. Although JG's reading had improved in the interval between the earlier investigation and this study, he continued to demonstrate the same pattern of performance under rapid presentation conditions, namely, significant effects of part of speech and imageability, and an inability to read nonwords. Whereas stimulation of the left hemisphere had no effect on his oral reading performance to briefly presented stimuli, application of stimulation to the homologous area on the right produced a significant decrement. JG read only 21% of words under right-sided transcranial magnetic stimulation, as opposed to 71% with no stimulation present.

Many cognitive neuropsychologists have remained sceptical of accounts that invoke right-hemisphere reading, at least in part on grounds of parsimony. Since the left hemisphere is competent in reading, why postulate a second set of reading mechanisms in the right hemisphere? In our view, the parsimony criterion is not as clear-cut as it might seem; an account that is judged unparsimonious for one data set may be more satisfying when it is applied to a wider range of phenomena. There are perhaps other ways to account for the evidence from alexia alone. When the right-hemisphere account of implicit reading is evaluated in the broader context of phenomena that involve processing by the right hemisphere, the parsimony argument is considerably weakened.

It cannot be denied, however, that the issue of right-hemisphere language representation remains a contentious one. This question has been addressed in a variety of studies, many of which have yielded inconsistent results and/or admit alternative interpretations. For example, some callosotomy patients have shown evidence of right-hemisphere language ability, whereas others have not (e.g. Baynes, 1990; Gazzaniga, 1983). In some cases, it has taken years for patients to demonstrate this capacity (e.g. Baynes, Wessinger, Fendrich, & Gazzaniga, 1995). On the other hand, the isolated right hemisphere of left-hemispherectomy patients generally manifests some language ability, which increases over time (e.g. Patterson et al., 1989; Smith, 1966; Vargha-Khadem et al., 1997). Additional evidence of right-hemisphere language capacity comes from studies of recovering aphasics, whose language performance often declines following a subsequent lesion to the right hemisphere (e.g. Basso, Gardelli, Grassi, & Mariotti, 1989; Papanicolaou, Moore, Deutsch, Levin, & Eisenberg, 1988). Right-hemisphere involvement in language recovery is also supported by recent functional imaging studies in aphasics, who show increased activity, relative to controls, in regions of the right hemisphere that are homotopic to language areas on the left (e.g. Cardebat et al., 1994; Weiller et al., 1995). And, although right-hemisphere lesions do not impair core language functions, they do affect performance on complex language tasks, such as those that involve discourse functions (see Beeman, 1997, for review). Evidence of right-hemisphere language ability in normal subjects has been particularly controversial. The criterion for right-hemisphere involvement is not simply that performance is worse when linguistic stimuli are delivered to the right, which could simply reflect degradation in the course of transferring information to the left hemisphere, but that the performance pattern is different. While some studies utilising split-field presentation have shown better left visual field performance for concrete than for abstract words, others have failed to demonstrate this pattern (see Patterson & Besner, 1984, for review). On the other hand, recent studies suggest quite different performance patterns for left and right visual field input on tasks that range from semantic priming (Beeman et al., 1994) to the resolution of lexical ambiguity (Burgess & Simpson, 1988; Faust & Gernsbacher, 1996).

Nevertheless, some researchers continue to dismiss the possibility that the right hemisphere contributes to language performance following damage to language areas in the left hemisphere. Rapp and Caramazza (1997) have argued, for example, that the language pattern attributed to the right hemisphere is unmotivated: Why should the right hemisphere be credited with orthographic and lexical/semantic capacities and not with phonological abilities? In response, we point out that in those cases where one can be reasonably confident that performance is mediated by the right hemisphere (as, for example, in the left visual field reading performance of Michel et al.'s [1996] patient with the posterior callosal lesion, or in the writing performance of a patient reported by Rapcsak, Beeson, and Rubens [1991], whose lesion destroyed virtually all of the left hemisphere), this is precisely the pattern that is observed. Why does the right hemisphere demonstrate this pattern? Perhaps because phonology is best developed in relation to spoken output, which appears to be strongly lateralised to the left hemisphere. The strongest objection to the right-hemisphere hypothesis is that it is untestable (e.g. Baynes, 1990; Rapp & Caramazza, 1997); this may no longer be a valid objection, given the new techniques for imaging and also for disrupting ongoing brain activity (e.g. transcranial magnetic stimulation). It is unlikely, however, that these data will tell a simple story. As the degree of language lateralisation seems to be at least partially under genetic control (viz., the apparent differences between right- and left-handers), the language capacity of the right hemisphere may vary. This may be one reason for the variability in callosotomy patients, and may explain why some alexics display implicit reading capacities while others do not. It is also possible that the acquisition as well as the expression of right-hemisphere language capacity varies with the distribution of attentional resources across the two hemispheres (e.g. Faure & Blanc-Garin, 1994).

The bi-hemispheric model is not the only account of the co-occurrence of LBL and covert reading in pure alexia. One class of hypotheses points to the very different requirements of the two types of reading tasks, in particular, that reading aloud requires conscious identification of a particular lexical item, whereas lexical decision and categorisation do not. On the assumption that pure alexia reflects damage to the left-hemisphere word form system (see Warrington & Shallice, 1980), Shallice and Saffran (1986, p. 452) suggested that weak activation of this system "could allow sufficient activation of the corresponding semantic representation to activate other representations by spreading activation, but not enough to inhibit the competing possibilities which explicit identification requires". Others, who attribute weak activation to problems in letter identification rather than to impairment of the word form system, have offered similar proposals (Arguin & Bub, 1993; Behrmann, Plaut, & Nelson, this issue). In this account, the system that mediates covert reading supports explicit word identification, via the LBL strategy, when the patient is allowed time to compensate for the letter identification problem. However, there are some covert readers (CB and EM in our series) who remain unable to identify letters no matter how much time they are given.

If weak activation does in fact stem from degraded orthographic input, one would expect alexics to show priming from letter strings that differ minimally from the target (e.g. *male* and *mele* should prime *mile*). Bowers, Arguin, and Bub (1996) tested this hypothesis in their patient IH. Although IH demonstrated priming from cross-case stimuli (e.g. *mile* primed *MILE*), he did not show priming effects of letter strings that differed from the target by only one letter. Other evidence from IH is also relevant here. Bowers, Bub, and Arguin (1996) demonstrated a word superiority effect with mixed case stimuli (e.g. CaMe) in this patient, which led them to conclude that his ability to identify letters was preserved. To account for the apparent preservation of orthographic processes in the context of letter-by-letter reading, Bowers, Arguin and Bub (1996, p. 561) have proposed that the reading pattern reflects a disconnection of orthographic codes from phonology: that is, letters and words can be identified but not phonologically encoded, and, further, that orthographic codes do not "interact with . . . semantic systems in a normal fashion".

This analysis seems remarkably close to the one we have suggested. Where we differ from Bowers and colleagues is in linking this disconnection to brain structure. We suggest that the reading pattern observed in pure alexia reflects orthographic access to the right-hemisphere lexicon. Right-hemisphere lexical access supports performance on implicit reading tasks, but does not map directly on to left-hemisphere semantic and phonological systems. To access these systems, patients who can do so resort to letter-by-letter reading.

Behrmann et al. (this issue) argue that the performance pattern of LBL readers across covert and explicit tasks reflects the processing of degraded orthographic input to the "same cascaded, interactive system that supported normal reading premorbidly" (p. 17). This system, they suggest, "probably involves both the left and the right hemispheres and . . . there is no reason to invoke the right-hemisphere reading system as the primary mediator of the lexical and semantic effects" (p. 46). We have no quarrel with the basic premise of Behrmann et al.'s analysis, namely, that the reading system is normally interactive, with top-down effects from semantics as well as bottom-up input from orthography, or with the possibility that the right hemisphere may contribute to normal reading (see papers in Beeman & Chiarello, 1997, for support for this idea). But, as argued earlier, we do see reason to implicate the right hemisphere, specifically, in pure alexics' performance on covert reading tasks, and the left hemisphere, specifically, in LBL reading. We would agree, however, that it is likely that the two sources of information can, and do, interact, for example in supporting the fast, partially recovered reading (with word class and imageability effects) that we have observed in some LBL readers.

useful information with respect to the capacities of the two hemispheres. We would make a similar argument for the analysis of the performance of LBL readers. In these patients, letter naming, and ultimately explicit identification of the word, entails the transfer of information from the right to the left hemisphere. It seems likely, moreover, that this information is in some way limited or degraded. What is the nature of the information that is passed from right to left? If the right hemisphere is indeed capable of extracting meaning from printed words, what is the nature of the information that is transmitted to left hemisphere? If the organisation of information in the two hemispheres differs—it has been claimed, for example, that right-hemisphere semantic information is "coarsely coded", in comparison with left-hemisphere semantics (e.g. Beeman et al., 1994)—how does highly processed input from one hemisphere intersect with differently encoded information in the other? These are basic questions for cognitive neuroscience, and critical questions to address in the continuing effort to understand the behaviour patterns of pure alexics.

CONCLUDING COMMENTS

Would it make sense to study cognition in commissurotomy patients without acknowledging the fact that the two hemispheres are disconnected? Of course not; recognition of this fact led to the development of new techniques for testing those patients and to much

REFERENCES

Albert, M.L., Yamadori., A., Gardner, H., & Howes, D. (1973). Comprehension in alexia. *Brain, 96,* 317–328.

Arguin, M., & Bub, D.N. (1993). Single-character processing in a case of pure alexia. *Neuropsychologia, 31,* 435–458.

Basso, A., Gardelli, M., Grassi, M.P., & Mariotti, M. (1989). The role of the right hemisphere in recovery from aphasia: Two case studies. *Cortex, 25,* 555–556.

Baynes, K. (1990). Language and reading in the right hemisphere: Highways or byways of the brain? *Journal of Cognitive Neuroscience, 2,* 159–179.

Baynes, K., Wessinger, C.M., Fendrich, R., & Gazzaniga, M.S. (1995). The emergence of the capacity to name left visual field stimuli in a callosotomy patient: Implications for functional plasticity. *Neuropsychologia, 33,* 1225–1242.

Beauregard, M., Chertkow, H., Bub, D., Murtha, S., Dixon, R., & Evans, A. (1997). The neural substrate for concrete, abstract, and emotional word lexica: A Positron Emission Tomography study. *Journal of Cognitive Neuroscience, 9,* 441–461.

Beeman, M. (1997). Coarse semantic coding and discourse comprehension. In M. Beeman & C. Chiarello (Eds.), *Right hemisphere language comprehension: Perspectives from cognitive neuroscience.* Mahwah, NJ: Lawrence Earlbaum Associates Inc.

Beeman, M., & Chiarello, C. (Eds.) (1997). *Right hemisphere language comprehension: Perspectives from cognitive neuroscience.* Mahwah, NJ: Lawrence Earlbaum Associates Inc.

Beeman, M., Friedman, R.B., Grafman, J., Perez, E., Diamond, S., & Lindsay, M.B. (1994). Summation priming and coarse semantic coding in the right hemisphere. *Journal of Cognitive Neuroscience, 6,* 26–45.

Behrmann, M., Black, S.E., & Bub, D. (1990). The evolution of pure alexia: A longitudinal study of recovery. *Brain and Language, 39,* 405–427.

Behrmann, M., Plaut, D.C., & Nelson, J. (this issue). A literature review and new data supporting an interactive account of letter-by-letter reading. *Cognitive Neuropsychology, 15*(1/2).

Behrmann, M., & Shallice, T. (1995). Pure alexia: An orthographic not spatial disorder. *Cognitive Neuropsychology, 12,* 409–454.

Binder, J.R., & Mohr, J.P. (1992). The topography of callosal reading pathways: A case control analysis. *Brain, 115,* 1807–1826.

Black, S.E., & Behrmann, M. (1994). Localization in alexia. In A. Kertesz (Ed.), *Localization and neuroimaging in neuropsychology* (pp. 331–376). San Diego, CA: Academic Press.

Bowers, J.S., Arguin, M., & Bub, D.N. (1996). Fast and specific access to orthographic knowledge in a case of letter-by-letter reading. *Cognitive Neuropsychology, 13,* 525–567.

Bowers, J.S., Bub, D.N., & Arguin, M. (1996). A characterisation of the word superiority effect in a case of letter-by-letter surface alexia. *Cognitive Neuropsychology, 13,* 415–442.

Bub, D.N., & Arguin, M. (1995). Visual word activation in pure alexia. *Brain and Language, 12,* 77–103.

Bub, D.N., Arguin, M., & Lecours, A.R. (1993). Jules Déjèrine and his interpretation of pure alexia. *Brain and Language, 45,* 532–559.

Bub, D.N., Black, S.E., & Howell, J. (1989). Word recognition and orthographic context effects in a letter-by-letter reader. *Brain and Language, 36,* 357–376.

Burgess, C., & Simpson, G. (1988). Hemispheric processing of ambiguous words. *Brain and Language, 33,* 86–104.

Buxbaum, L., & Coslett, H.B. (1996). Deep dyslexic phenomena in a pure alexic. *Brain and Language, 54,* 136–167.

Caplan, L.R., & Hedley-Whyte, T. (1974). Cueing and memory dysfunction in alexia without agraphia: A case report. *Brain, 97,* 251–262.

Cardebat, D., Demonet, J.-F., Celsis, P., Puel, M., Viallard, G., & Marc-Vergnes, J.-P. (1994). Right temporal compensatory mechanisms in a deep dysphasic patient: A case report with activation study by PET. *Neuropsychologia, 32,* 97–103.

Cohen, L., & Dehaene, S. (1995). Number processing in pure alexia: The effect of hemispheric asymmetries and task demands. *Neurocase, 1,* 121–138.

Coltheart, M. (1980). Deep dyslexia: A right hemisphere hypothesis. In M. Coltheart, K. Patterson, & J.C. Marshall (Eds.), *Deep dyslexia* (pp. 326–380). London: Routledge.

Coslett, H.B., & Monsul, N. (1994). Reading and the right hemisphere: Evidence from transcranial magnetic stimulation. *Brain and Language, 46,* 198–211.

Coslett, H.B., & Saffran, E.M. (1989a). Evidence for preserved reading in pure alexia. *Brain, 112*, 327–329.

Coslett, H.B., & Saffran, E.M. (1989b). Preserved object recognition and reading comprehension in optic aphasia. *Brain, 112*, 1091–1110.

Coslett, H.B., & Saffran, E.M. (1992). Optic aphasia and the right hemisphere: A replication and extension. *Brain and Language, 43*, 148–161.

Coslett, H.B., & Saffran, E.M. (1994). Mechanisms of implicit reading in pure alexia. In M.J. Farah & G. Ratcliff (Eds.), *The neuropsychology of high-level vision*. Hillsdale, NJ: Lawrence Erlbaum Associates Inc.

Coslett, H.B., Saffran, E.M., Greenbaum, S., & Schwartz, H. (1993). Reading in pure alexia: The effect of strategy. *Brain, 116*, 21–27.

Cossu, G., da Prati, E., & Marshall, J.C. (1995). Deep dyslexia and the right-hemisphere hypothesis: Spoken and written language after extensive left-hemisphere lesion in a 12-year-old boy. *Cognitive Neuropsychology, 12*, 391–407.

Damasio, A., & Damasio, H. (1983). The anatomic basis of pure alexia. *Neurology, 33*, 1573–1583.

Déjèrine, J. (1892). Contribution a l'étude anatomo-pathologique et clinique des différentes variétés de cécité verbale. *Comptes Rendu des Seances de la Societé Biologique, 4*, 61–90.

Ellis, A.W., Young, A.W., & Anderson, C. (1988). Modes of word recognition in the left and right cerebral hemispheres. *Brain and Language, 35*, 254–273.

Farah, M.J., & Wallace, M. (1991). Pure alexia as a visual impairment: A reconsideration. *Cognitive Neuropsychology, 8*, 313–334.

Faure, S., & Blanc-Garin, J. (1994). Right hemisphere semantic performance and competence in a case of partial interhemispheric disconnection. *Brain and Language, 47*, 557–581.

Faust, M.E., & Gernsbacher, M.A. (1996). Cerebral mechanisms for supression of inappropriate information during sentence comprehension. *Brain and Language, 53*, 234–259.

Feinberg, T., Dyches-Berke, D., Miner, C.R., & Roane, D.M. (1994). Knowledge, implicit knowledge and metaknowledge in visual agnosia and pure alexia. *Brain, 118*, 789–800.

Friedman, R.B., & Alexander, M.P. (1984). Pictures, images and pure alexia: A case study. *Cognitive Neuropsychology, 1*, 9–23.

Friedman, R.B., & Hadley, J.A. (1992). Letter-by-letter surface alexia. *Cognitive Neuropsychology, 9*, 185–208.

Gazzaniga, M.S. (1983). Right hemisphere language following bisection: A 20-year perspective. *American Psychologist, 38*, 525–537.

Geschwind, N. (1965). Disconnexion syndromes in animals and man: Part II. *Brain, 88*, 585–644.

Grossi, D., Fragassi, N.A., Orsini, A., De Falco, F.A., & Sepe, O. (1984). Residual reading capability in a patient with alexia without agraphia. *Brain and Language, 23*, 337–348.

Howard, D. (1991). Letter-by-letter readers: Evidence for parallel processing. In D. Besner & G. Humphreys (Eds.), *Basic processes in reading* (pp. 34–76). Hillsdale, NJ: Lawrence Erlbaum Associates Inc.

Kay, J., & Hanley, R. (1991). Simultaneous form perception and serial letter recognition in a case of letter-by-letter reading. *Cognitive Neuropsychology, 8*, 249–273.

Kinsbourne, M., & Warrington, E.K. (1962). A disorder of simultaneous form perception. *Brain, 85*, 461–486.

Kreindler, A., & Ionnescu, Y. (1961). A case of "pure" word blindness. *Journal of Neurology, Neurosurgery and Psychiatry, 24*, 275–280.

Landis, T., Regard, M., & Serrat, A. (1980). Iconic reading in a case of alexia without agraphia caused by a brain tumour: A tachistoscopic study. *Brain and Language, 11*, 45–53.

Levine, D.M., & Calvanio, R.A. (1978). A study of the visual defect in verbal alexia-simultanagnosia. *Brain, 101*, 65–81.

Lhermitte, F., & Beauvois, M.-F. (1973). A visual-speech disconnexion syndrome: Report of a case with optic aphasia, agnoic alexia and colour agnosia. *Brain, 96*, 695–714.

Michel, F., Henaff, M.A., & Intrilligator, J. (1996). Two different readers in the same brain after a posterior callosal lesion. *NeuroReport, 7*, 786–788.

Pandya, D.N., & Seltzer, B. (1986). The topography of commissural fibers. In F. Lepore, M. Ptito, & H.H. Jasper (Eds.), *Two hemispheres—one brain*. New York: Liss.

Papanicolaou, A.C., Moore, B., Deutsch, G., Levin, H.S., & Eisenberg, H.M. (1988). Evidence for right-hemisphere involvement in recovery from aphasia. *Archives of Neurology, 45*, 1025–1029.

Patterson, K., & Besner, D. (1984). Is the right hemisphere literate? *Cognitive Neuropsychology, 1*, 315–341.

Patterson, K., & Kay, J. (1982). Letter-by-letter reading: Psychological descriptions of a neurological syndrome. *Quarterly Journal of Experimental Psychology, 34A*, 411–441.

Patterson, K., Vargha-Khadem, F., & Polkey, C.F. (1989). Reading with one hemisphere. *Brain, 112*, 39–63.

Rapcsak, S.Z., Beeson, P.M., & Rubens, A.B. (1991). Writing with the right hemisphere. *Brain and Language, 41*, 510–530.

Rapp, B., & Caramazza, A. (1997). The modality-specific organization of grammatical categories: Evidence from impaired spoken and written sentence production. *Brain and Language, 56*, 248–286.

Reuter-Lorenz, P.A., & Baynes, K. (1992). Modes of lexical access in the callosotomized brain. *Journal of Cognitive Neuroscience, 4*, 155–164.

Reuter-Lorenz, P., & Brunn, J. (1990). A prelexical basis for letter-by-letter reading: A case study. *Cognitive Neuropsychology, 7*, 1–20.

Rizzo, M., & Robin, D.A. (1996). Bilateral effects of unilateral visual cortex lesions in human. *Brain, 119*, 951–963.

Saffran, E.M., Bogyo, L.C., Schwartz, M.F., & Marin, O.S.M. (1980). Does deep dyslexia reflect right hemisphere reading? In M. Coltheart, K.E. Patterson, & J.C. Marshall (Eds.), *Deep dyslexia*. London: Routledge.

Schweiger, A., Zaidel, E., Field, T., & Dobkin, B. (1989). Right hemisphere contribution to lexical access in an aphasic with deep dyslexia. *Brain and Language, 37*, 73–89.

Sekuler, E.B., & Behrmann, M. (1996). Perceptual cues in pure alexia. *Cognitive Neuropsychology, 13*, 941–974.

Shallice, T., & Saffran, E.M. (1986). Lexical processing in the absence of explicit word identification: Evidence from a letter-by-letter reader. *Cognitive Neuropsychology, 3*, 429–458.

Smith, A. (1966). Speech and other functions after left (dominant) hemispherectomy. *Journal of Neurology, Neurosurgery and Psychiatry, 29*, 467–471.

Vargha-Khadem, F., Carr, L.J., Isaacs, E., Brett, F., Adams, C., & Mishkin, M. (1997). Onset of speech after left hemispherectomy in a 9-year old boy. *Brain, 120*, 159–182.

Warrington, E.K., & Shallice, T. (1980). Word-form dyslexia. *Brain, 103*, 99–112.

Weiller, C., Isensee, C., Rijntjes, M., Huber, W., Muller, S., Bier, D., Dutschka, K., Woods, R.P., Noth, J., & Diener, H.C. (1995). Recovery from Wernicke's aphasia: A positron emission tomographic study. *Annals of Neurology, 37*, 723–732.

Young, A.W., & Ellis, A.W. (1985). Different methods of lexical access for words presented in the left and right visual hemifields. *Brain and Language, 24*, 326–358.

Zaidel, E., & Peters, A.M. (1981). Phonological encoding and ideographic reading by the disconnected right hemisphere: Two case studies. *Brain and Language, 14*, 205–234.

Perceptual and Lexical Factors in a Case of Letter-by-Letter Reading

Doriana Chialant
University of Massachusetts at Amherst, and Tewksbury Hospital, MA, USA

Alfonso Caramazza
Harvard University, Cambridge, USA

We report the case of a letter-by-letter reader (MJ) who showed normal processing of single letters and who could normally access the orthographic input lexicon when presented with letter names for aural recognition, or when allowed enough time to process a visually presented letter string. However, MJ showed severe difficulties in simultaneously processing multiple letters and other simple visual stimuli. Furthermore, she does not have normal access to lexical orthographic representations and their meanings when stimuli are presented for too brief a time to allow for serial processing of the letter string. We found no evidence of (partial) lexical or semantic access without corresponding recognition of the letters in a word: No signs of implicit reading were observed when the input stimuli were controlled for the relevant visual features; "implicit reading" was only obtained under conditions that allowed sophisticated guessing. This pattern of results is interpreted as indicating that LBL reading (at least in MJ) results from damage to prelexical processing mechanisms. In MJ's case, the deficit reflects the degraded transfer of information from a normal visual processing system in the right hemisphere to a normal language processing system in the left hemisphere.

Requests for reprints should be addressed to Alfonso Caramazza, Department of Psychology, Harvard University, 33 Kirkland Street, Cambridge, MA 02138, USA. Email: caram@wjh.harvard.edu

The work reported here was supported by NIH grant NS 34073 to Harvard University. This paper is dedicated to MJ, who in spite of her difficulties remaining seated for prolonged periods of time, and of her aversion to computers, remained cooperative and in good spirit throughout the testing phase of this project. She also taught the first author pretty much all she knows about lobsters. The authors would also like to thank JG and TG for their participation in the study, Michele Miozzo for his comments on various aspects of the project, and Kathryn Link, Max Coltheart, and two anonymous referees for their comments on an earlier version of this paper.

INTRODUCTION

What is the nature of the damage such that reading becomes laborious, marked by a consistent word length effect, and apparently characterised by a letter-by-letter, serial processing of the input?

Letter-by-letter (LBL) reading, also known as "pure alexia[1]" or "alexia without agraphia," is a form of acquired dyslexia in which patients appear to read single words by identifying them one letter at a time, with the result that reading is characterised by a marked word length effect. There is great variability in the extent to which reading is slowed down in different patients (cf. case JG in Coslett & Saffran, 1989; case KC in Patterson & Kay, 1982; case BL in Friedman & Hadley, 1992; case RAV in Warrington & Shallice, 1980). This deficit occurs in the face of sparing of all other linguistic modalities and, in particular, of the ability to recognise orally spelled words and of the ability to write.

From a traditional neurological standpoint, LBL reading has been interpreted as a disconnection syndrome in which a right visual field cut prevents direct access to information in the left hemisphere, and damage to the splenium of the callosum alters interhemispheric transfer (Damasio & Damasio, 1986, Déjèrine, 1982; Geschwind, 1965; Greenblatt, 1973).

Since Déjèrine's (1892) first interpretation of this phenomenon as a disconnection syndrome between the occipital cortices and the left hemisphere's mechanisms that allow word recognition, LBL reading has been interpreted functionally in numerous ways. There are at least three distinct classes of interpretations. One class of interpretations views LBL reading as a low-level perceptual deficit that affects visual processing for linguistic and nonlinguistic stimuli alike (Friedman & Alexander, 1984; Rapp & Caramazza, 1991). A second class of interpretations views LBL reading as a deficit in letter recognition (Arguin & Bub, 1994, Howard, 1990). Finally, a third class of interpretations views LBL reading as a deficit at the word level (Bub, Black & Howell, 1989; Shallice & Saffran, 1986; Warrington & Shallice, 1980).

Patterson and Kay (1982) compared information from several clinical studies of LBL readers and concluded that the phenemenon is independent of: (a) hemianopsia, since there are LBL readers without visual field cuts, and hemianopic cases with reading impairments other than LBL reading (Greenblatt, 1973; Patterson & Kay, 1982); (b) simultanagnosia, since processing in free vision of complex pictures seems unaffected; (c) attention and short-term visual memory deficits, since these functions were good in at least some cases (Warrington & Shallice, 1980); and (d) decreased visual acuity, as this function was shown to be normal in a number of cases (Warrington & Shallice, 1980). The authors thus concluded that LBL reading does not result from a low-level visual

---

[1] Pure alexia is a broader category of reading disorders than LBL reading. It includes those cases in which patients are virtually incapable of reading any word (see Miozzo & Caramazza, this issue).

perceptual deficit. Later findings of implicit reading in some LBL cases (Coslett & Saffran, 1989, 1994a, 1994b) have also been interpreted as evidence against the perceptual deficit accounting of LBL reading.

More recently, testing of perceptual processing with limited exposures has shown that in some patients the deficit can extend to stimuli other than words, and may be affected by spatial and perceptual properties of the input (Farah & Wallace, 1991; Kay & Hanley, 1991; Rapp & Caramazza, 1991). For example, some investigators have found that visual processing of complex pictures presented at brief exposures is slowed down in these patients in comparison to normal subjects (Friedman & Alexander, 1984), and that visual processing of items presented in arrays is also impaired under limited visual exposure conditions (Farah & Wallace, 1991; Kinsbourne & Warrington, 1962; Levine & Calvanio, 1978; Rapp & Caramazza, 1991).

Evidence of impaired single letter processing, in at least some LBL readers, has been invoked as an account of this reading deficit (Arguin & Bub, 1994; Howard, 1990). Arguin and Bub (1994), in a series of elegant experiments, demonstrated: (1) that their patient showed letter recognition difficulties in the presence of preserved processing of structural representations, and (2) that this letter recognition disorder was due to a loss of "automatic" identification of familiar characters. However, as pointed out by the authors themselves, evidence demonstrating impaired letter processing in these patients does not rule out impairments at other levels of processing. Furthermore, it remains unclear why on this account LBL readers do not produce many more visual errors in reading (e.g. bear → pear; bear → beer).

Warrington and Shallice named the LBL reading deficit "word form dyslexia" to reflect their hypothesis that the damage responsible for the disorder is at the level of orthographic lexical forms. A consequence of this type of damage is to force patients to process words sublexically by naming the letters and accessing the lexicon through the intact spelling system[2]. Variants of Warrington and Shallice's interpretation of LBL as resulting from an impairment to the word form system have been adopted in modified form by others (Bowers, Arguin, & Bub, 1996a, Bowers, Bub, & Arguin, 1996b; Patterson & Kay, 1982). Patterson and Kay (1982) proposed a revised interpretation according to which damage occurs not to the word form system itself, but rather to the access procedures to this system. This induces a letter-by-letter parsing of the input string, with the consequence that whole-word recognition is not possible. These versions of the "word form dyslexia" interpretation of LBL reading have been challenged by the finding that some patients with LBL reading show word superiority effects (WSE) when lexical and nonlexical items are shown at exposures too brief for overt recognition and report (Bub et al., 1989;

---

[2] No explicit account has ever been provided of how this process is supposed to work.

Reuter-Lorenz & Brun, 1990; but see Kay & Hanley, 1991). To accommodate the latter results, Bowers et al. (1996a, 1996b) propose instead that LBL reading is the result of a disconnection between a normally functioning orthographic input lexicon and normally functioning semantic and phonological lexical components. However, all three "word form dyslexia" accounts of LBL reading encounter difficulties in explaining the finding that some LBL readers show (partial) semantic access (as evidenced by above-chance performance on categorisation tasks) of lexical items that were shown at exposures too short for overt reading and recognition (Coslett & Saffran, 1989; Coslett, Saffran, Greenbaum, & Schwartz, 1993; Shallice & Saffran, 1986).

An entirely different approach has been taken by Coslett and Saffran (1989). They proposed that the word superiority and categorisation effects in LBL readers arise not in the left-hemisphere word recognition system (which is supposedly either damaged or accessed in a letter-by-letter fashion) but in a functionally and anatomically independent lexical system located in the right hemisphere. On this interpretation, LBL reading can be used to investigate the processing structure of reading mechanisms in the right hemisphere. Coslett and Saffran argue that LBL readers' performance is best understood in the following way: Whenever explicit word identification is required, visual input is first analysed in the right hemisphere and information is transmitted in a serial, laborious fashion to the intact left-hemisphere word form system; however, when explicit identification is not required, then the visual input is processed by the right hemisphere word recognition system (which does not allow overt identification because of its lack of a phonological output system).

Although the various accounts of LBL reading briefly reviewed here are often discussed as alternative explanations of the disorder (e.g. Behrmann & Shallice, 1995), it is entirely possible that they each constitute plausible explanations for *different* forms of a heterogenous reading disorder which share only the superficial characteristic of LBL reading. It is important, therefore, to describe clearly the performance of such cases on the relevant dimensions along which supposedly alternative accounts can be compared. In this paper, we report the case of MJ, a 62-year-old woman, who suffered a left occipital infarct in 1989 resulting in a complete right homonymous hemianopia, and in laborious reading with characteristic features of LBL reading. Here, we attempt to (1) evaluate whether MJ's reading performance satisfies the criteria for being considered a LBL reader, (2) explore whether MJ's reading difficulty is related to her visual field cut, (3) explore whether her visual processing difficulty is limited to verbal materials, (4) investigate whether she demonstrates implicit reading in categorization tasks, and (5) explore whether this case can further our understanding of implicit reading processes. Her performance is compared to that of a hemianopic patient (JG) who does not show LBL reading and to that of a neurologically intact control subject (TG).

CASE HISTORIES

MJ, a right-handed woman, was 62 years old at the time of testing. She is a high-school graduate and currently a homemaker. Premorbidly she was an avid reader. MJ suffered a left occipital CVA on 27 June 1989, which left her with right-side weakness, a right visual field cut, and severe reading difficulties. Perimeter tests of her visual fields dated January 1995 reveal a complete right visual field cut with no macular sparing. A brain MRI dated 3 July 1989 shows a 1.5cm lesion in the left optic radiation near the thalamus along with multiple periventricular white matter lesions with involvement of the corpus callosum.

Language testing revealed normal processing in all tasks but reading. MJ's speech was fluent and normally formed, with no evidence of dysarthria. Word and sentence level processing were normal. MJ performed flawlessly on all comprehension (grammaticality judgement; picture-sentence/word matching), naming (oral and written naming to picture and tactile confrontation), and repetition tasks. Written spelling, oral spelling, and aural recognition of orally spelled words were also accurate (98%, 95%, and 96% correct, respectively). Performance on visual tasks with unlimited exposure was flawless (drawing, copying with and without delay, line cancellation and bisection, search tasks) (see Fig. 1), as was letter naming at 100msec (upper and lower case) and 16msec exposures (upper case) (see Table 1). Colour naming and face recognition were normal.

Table 1. MJ's Letter Naming Performance in Free Vision and at 16msec Exposure

Exposure	Case	%	No
1sec	Upper	100	(26/26)
	Lower	100	(26/26)
16msec	Upper	96	(96/100)

Errors: Q > O; W > V; B > D; G > C.

Reading performance, although accurate (94% correct) and unaffected by lexical factors (concreteness, word category, regularity, and frequency) (see Table 2), was characterised by LBL naming prior to word reading. Sentence reading was virtually impossible. At limited

Table 2. MJ's accuracy (% Correct) in Reading Single Words at Unlimited and at 500msec Exposures

	Exposures	
Lexical Effects	Unlimited	500msec
Abstractness		
Abstract	100	40
Concrete	100	40
Grammatical class		
Nouns	86	57
Verbs	93	30
Adjectives	86	47
Functors	90	70
Spelling-to-sound regularity		
Regular consistent	100	50
Regular inconsistent	97	50
Exceptional	90	50
Frequency		
High frequency	96	60
Low frequency	91	34
Total	94	48

Fig. 1. MJ's line cancellation, line bisection, and copying performance.

Table 3. Error Percentage per Letter Position in Reading for MJ (500 msec Exposure) and JG (Free Vision)

	N	Word Length	1	2	3	4	5	6
MJ	51	4	8	9	31	96		
	20	5	5	25	45	95	110	
	38	6	3	15	45	79	92	103
JG	35	4	11	16	33	73		
	34	5	3	6	15	47	85	
	28	6	4	7	22	29	64	72

exposure (500msec), MJ's reading performance markedly deteriorated and a clear frequency effect emerged (see Table 2). Her reading errors were mainly lexical substitutions that tended to share the initial letters with the target word (e.g. belief → below). Error analysis revealed that the percentage of errors per letter position is a function of distance from the beginning of the word (see Table 3).

JG is a right-handed, 66-year-old woman, with 2 years of college. JG was also an avid reader premorbidly. She suffered a first left posterior occipital CVA on February 1994, and a second CVA in April, 1994, which involved the left temporal-occipital region. Brain MRI revealed a focal abnormality in the occipital area, with mild bilateral involvement. JG presents with a right visual field cut, and right-side weakness. She reports experiencing a right visual field cut on both the upper and lower quadrants, which was confirmed during evaluation. JG also complains of some memory loss as well as difficulties in calculation.

Performance on language screening tests revealed moderately decreased reading accuracy with visual/morphological errors, a mild word finding difficulty in confrontation naming tests with semantic errors, and mildly impaired spelling and number processing abilities. Performance on visual tasks was normal except for delayed copying and for face recognition which were mildly impaired. Colour naming was flawless.

In free vision, JG's reading performance was impaired. Her reading errors, like MJ's, were mainly lexical substitutions that tended to share the initial letters with the target word. Error analysis revealed that, also like MJ, JG's percentage of errors per letter position tended to be a function of distance from the beginning of the word (see Table 3).

TG is a right-handed, 59-year-old woman with high-school education. She worked as a proofreader. TG participated in the study as a control subject. Her medical history is not significant for brain damage.

IS MJ A LETTER-BY-LETTER READER?

A set of tasks was used to investigate MJ's ability to process written language and rule out gross visual perceptual deficits. The tasks included reading in free vision and at different exposure durations, lexical decision tasks, aural recognition of spelling words, and oral and written spelling tasks. An object naming task was included to assess gross perceptual processing.

Reading in Free Vision

MJ was administered a reading task comprising 120 words ranging in length from 3 to 8 letters (20 words of each length), matched for frequency and concreteness. Words were presented one at a time, printed in 24-point capital letters, centred on a sheet of paper. The words were first presented displayed horizontally, and later, at regular intervals of 1 month, they were presented displayed vertically and spaced out horizontally. Each display type presented the words in a different random order. Different display types were used to rule out the possibility that increases in reading latencies for letters on the rightmost side of words were due to a mild form of visual neglect. Reading latencies were measured with a stop-watch from the time the word was presented to the time a lexical response was given. Only correct responses were included in the length effect analyses.

There was no effect of word length on MJ's reading accuracy (see Table 4a), and errors (10, 12, and 11 responses for the horizontal, spaced, and vertical presentations, respectively) were mainly lexical substitutions that shared the initial letters with the target word (e.g. symphony → sympathy). A two-way ANOVA was performed on correct response latencies. The results showed a main effect of display types [$F(2,289) = 4.8$, $P < .01$], and of length [$F(5,289) = 28.5$, $P < .0001$] but no significant interaction [$F(10,289) = 1.1$, $P < .4$]. This indi-

Table 4a. MJ's Reading Accuracy (% and No. Correct) for Topographically Different Word Displays at Unlimited Exposure

Word Length	Horizontal	Spaced	Vertical
3	89 (16/18)	90 (18/20)	100 (19/19)
4	95 (19/20)	100 (19/19)	81 (13/16)
5	95 (19/20)	85 (17/20)	90 (17/19)
6	100 (18/18)	95 (18/19)	90 (17/19)
7	88 (14/16)	78 (14/18)	90 (17/19)
8	80 (16/20)	90 (18/20)	90 (18/20)
Total	91 (102/112)	90 (104/116)	90 (101/112)

e.g. Errors: Prestige → precise; circuit → circle; symphony → sympathy; bit → bill.

cates that MJ's reading performance is affected by a word length effect that is independent of display type (see Table 4b).

Mean response latencies and standard deviations for the different displays are shown in Table 4b. The comparison among the three subjects for the horizontal condition is shown in Fig. 2. MJ's reading performance for horizontally displayed words was compared with that of JG and TG. An overall two-way analysis of variance indicates a main effect of subject [$F(2,304) = 11.9$, $P < .0001$], a main effect of length [$F(5,304) = 1.4$, $P < .0001$], and a significant interaction [$F(10,304) = 10.6$, $P < .0001$]. Individual, two-way ANOVAs showed the following: For MJ versus JG there was a significant effect of subject [$F(1,194) = 1.1$, $P < .0001$], a significant effect of length [$F(5,194) = 10.7$, $P < .0001$], and a significant interaction [$F(5,194) = 9.8$, $P < .0001$]; for MJ versus TG there was a significant effect of subject [$F(1,206) = 1.7$, $P < .0001$], a significant effect of length [$F(5,206) = 11.9$, $P < .0001$], and a significant interaction [$F(5,206) = 11.7$, $P < .0001$]; and for JG versus TG there was a main effect of subject [$F(1,208) = 2$, $P < .0001$], no effect of length [$F(5,208) = 2.1$, $P < .07$], and no significant interaction [$F(5,208) = 2.1$, $P < .08$]. To assess further the effects of word length on reading performance, a series of independent t-tests was carried out for each subject, comparing performance at each length with performance at every other length. The comparisons for MJ revealed significant differences for each comparison except for five-letter vs. six-letter [$t(35) = 0.7$, $P < .5$] and seven-letter vs. eight-letter lengths [$t(28) = 0.09$, $P < .1$]; for JG no significant differences were obtained at any length, except for the shortest length (three-letter words), which was significantly different from lengths five-letter [$t(34) = -2.4$, $P < .02$], six-letter [$t(32) = -3.1$, $P < .004$], seven-letter [$t(30) = -2.8$, $P < .01$], and eight-letter [$t(32) = -2.3$, $P < .03$]; and for TG, no significant difference was obtained for any comparison. The results indicate that MJ's performance is qualitatively different from that of JG and TG. In particular, MJ's performance for horizontally displayed words is characterised by a marked length effect, with stepwise increases in reading latencies of up to several seconds. JG's performance for horizontal displays, although slower overall than that of the normal control, is not characterised by such progressive increases in latencies with longer stimuli.

Table 4b. MJ's Mean Response Time (Sec) and SD for Topographically Different Word Displays at Unlimited Exposure

Word Length	Horizontal[a]	Spaced	Vertical
3	1.4 (0.8)	1.6 (0.8)	2.2 (1.1)
4	2.5 (1.1)	3.0 (1.8)	3.5 (2.7)
5	4.5 (3.7)	3.6 (1.8)	4.2 (2.4)
6	3.8 (1.9)	3.4 (1.8)	6.1 (4.3)
7	7.3 (4.9)	5.1 (2.4)	8.3 (3.8)
8	7.5 (4.4)	7.5 (4.2)	7.6 (4.4)

[a]No effect of imageability on accuracy or RT; frequency effect on RT but not on accuracy.

Lexical Decision

In order to rule out the possibility that output processing deficits may be contributing to MJ's

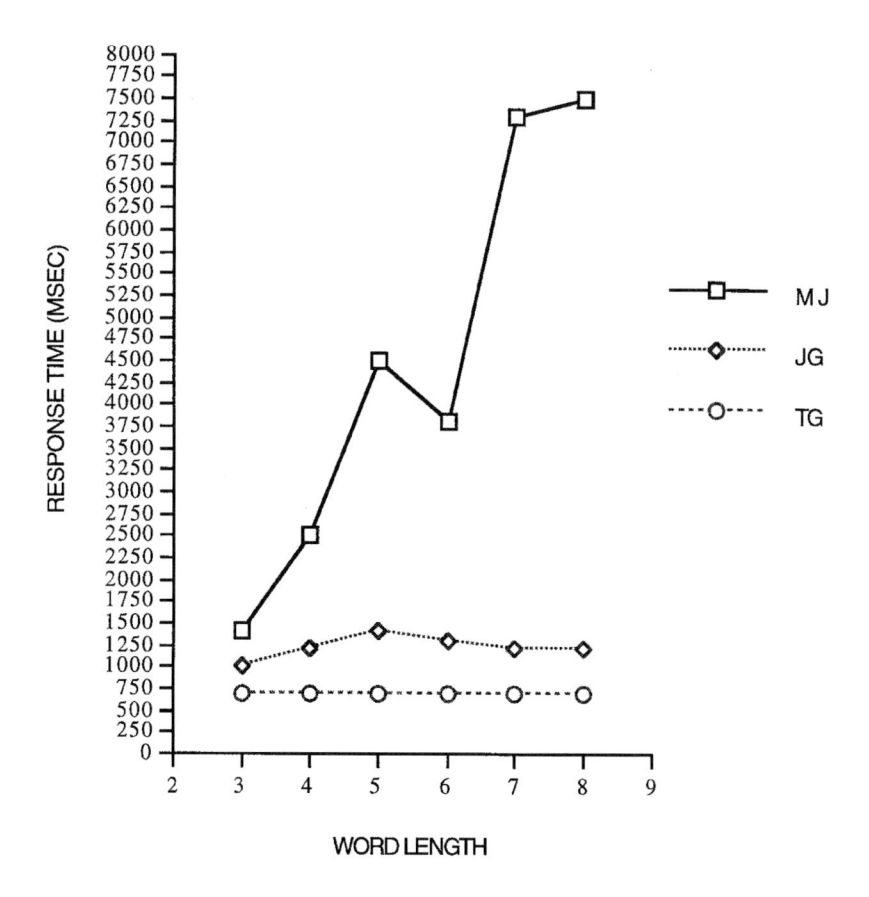

Fig. 2. MJ, JG, and TG—reading performance for horizontal displays.

reading difficulties, MJ was asked to perform a lexical decision task[3].

MJ was shown words and nonwords on a computer screen. Words were either four, six, or eight letters long and were matched for frequency. Nonwords were length-matched to the words and respected orthographic constraints of English orthography. There were 10 stimuli per experimental condition. Words and nonwords of different lengths were randomly intermixed. The following procedure was used: A fixation point would appear in the middle of the screen, followed by a 17msec screen refresh time and then by a target item printed in 24-point capital letters. The stimulus remained on the screen until a response was produced. Response times were measured

---

[3] As JG's lack of length effect in reading was considered sufficient evidence to conclude that she was not a LBL reader, and due to time limitations, JG was not administered the tasks in experiments 2 and 3.

from stimulus onset to response. MJ was instructed to look at the screen and press a computer key labelled "Y" for "yes" when she thought the stimulus was a word, and a key labelled "N" for "no" when she thought the stimulus was not a word. Stimuli were presented in two blocks. For the first block, the "Y" key was on MJ's left and the "N" key was on her right. For the second block, the locations of the "yes" and "no" keys were reversed. Only correct responses (Table 5) were included in the analysis of reaction times (Fig. 3).

MJ's overall accuracy was high (97% for both words and nonwords), but reaction times showed an effect of stimulus length [$F(2,32) = 7.8$, $P < .002$]: Longer items elicited longer reaction times, with abnormally steep increases. There was also a main effect of stimulus type, with words being recognised consistently faster than nonwords [$F(1,32) = 11.2$, $P < .004$]. There was no significant interaction between stimulus type and length [$F(2,32) = 0.5$, $P < .6$]. The presence of a main effect of stimulus length on reaction times in a task that only requires visual recognition, but no spoken output, indicates that MJ's abnormal reading latencies are most likely to be due to an input processing deficit and not to a deficit in accessing a phonological representation for spoken output.

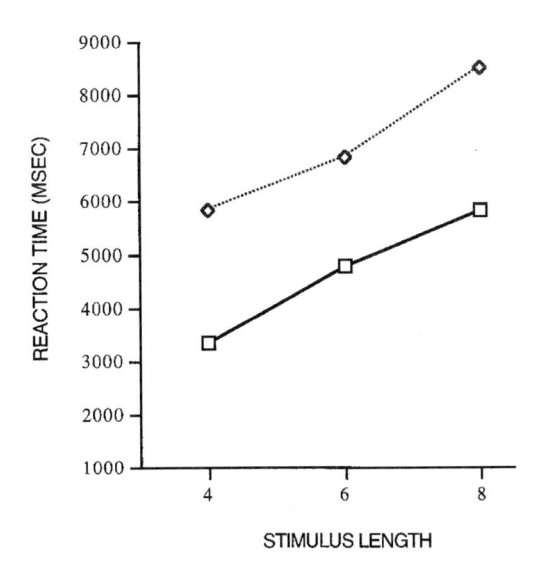

Fig. 3. MJ—lexical decision latencies.

Table 5. MJ's Lexical Decision Accuracy (% and No. Correct) for Words and Nonwords

	Stimuli	
Word Length	Words	Nonwords
4	100 (10/10)	90 (9/10)
5	90 (9/10)	100 (10/10)
6	100 (10/10)	100 (10/10)
Total	97 (29/30)	97 (29/30)

Reading Accuracy at Different Exposure Durations

MJ was shown words of different lengths at different exposure durations. Three sets of words, matched for frequency and spelling-to-sound regularity, were used. Each set was presented at a different exposure duration—400msec, 500msec, and unlimited exposure, respectively. Within each set, four-, six-, and eight-letter words were randomly intermixed. Stimuli were presented centred on a computer screen, but no attempt was made to control eye movements. Stimuli were shown in 24-point capital letters. MJ was asked to read the words aloud. Her performance is summarised in Table 6 and Fig. 4.

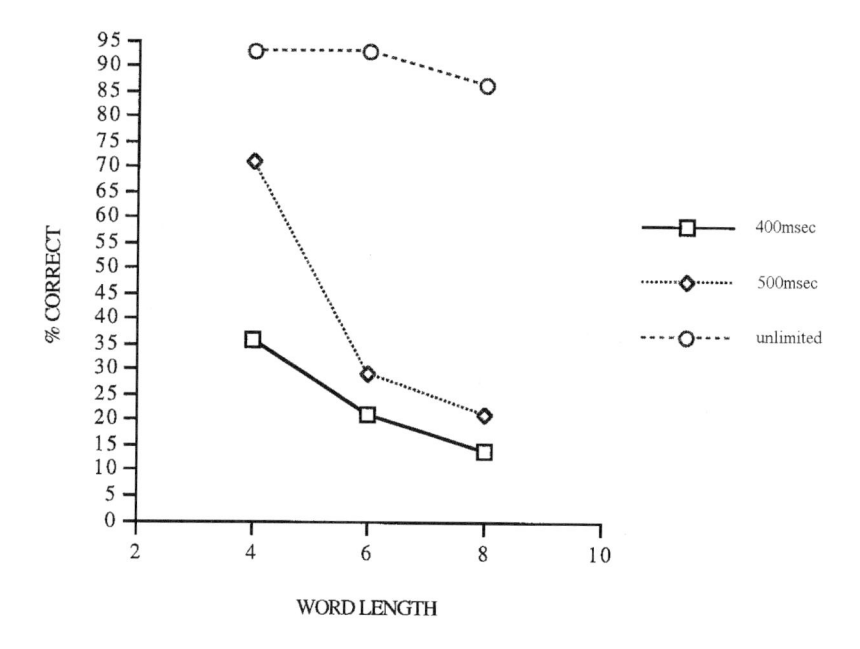

Fig. 4. MJ—The relationship between word length and reading accuracy at different exposure durations.

MJ's performance shows a marked effect of length [Friedman $\chi r^2(2) = 6, P < .05$]. Exposure duration greatly affected performance: Brief exposures resulted in very poor reading accuracy, with poorer performance for longer words. These results indicate that MJ's reading accuracy is sensitive to the manipulation of visual variables such as exposure duration. In particular, they indicate that at brief exposures MJ does not have time to scan the visual input letter by letter, and she is forced to "guess" on the basis of whatever partial letter information she can extract from the stimulus (usually parts of the beginning of the word).

Table 6. MJ's Reading Accuracy at Different Exposure Durations for Words of Different Length—% (No.) Correct

	Exposure		
Word Length	400	500	Unlimited
4	36 (5/14)	71 (10/14)	93 (13/14)
6	21 (3/14)	29 (4/14)	93 (13/14)
8	14 (2/14)	21 (3/14)	86 (12/14)
Total	24 (10/42)	41 (17/42)	91 (38/42)

Recognition of Spelled Words and Oral and Written Spelling Production

Results of the reading and lexical decision tasks clearly indicate that MJ has severe difficulties in recognising written words. A possible cause of these difficulties could be damage to her knowledge of the orthographic structure of words. To assess the integrity of this knowledge, MJ was administered the following three tasks: aural recognition of spelled words, oral spelling, and writing to dictation. The same list of words used in the reading test were used in these tasks. Words were presented in different random orders in each task, which were administered at 1-month intervals from the previous task. In the aural recognition task, the examiner spelled aloud a word and MJ was asked to identify it (e.g. "bee", "ay", "tee" for bat). In the oral and written spelling tasks, the examiner said a word aloud and MJ was required to spell it either orally or in writing. Her accuracy on these tasks is summarised in Table 7. MJ performed these tasks quickly and without difficulty. There was no effect of letter position on the very few errors produced in these tasks. These results indicate that MJ's difficulties in reading are not due to a general problem in processing orthographic representations. Furthermore, they indicate that her difficulties in processing orthographic representations occur prior to the point at which graphemic strings are used to access lexical orthographic forms. For if this were not the case, then MJ's performance on the aural recognition task would have been characterised by the same difficulties she encountered in reading.

Object Naming

Object naming is an extremely complex task involving many cognitive and linguistic components. Poor performance on this task is hardly informative since there are many possible causes that could be responsible for the performance. However, good performance would suggest that the component processes, from visual recognition to speech articulation,

Table 7. MJ's Aural Recognition and Spelling Accuracy (% and No. Correct)

Word Length	Aural Recognition	Oral Spelling	Writing to Dictation
3	100 (20/20)	100 (20/20)	100 (20/20)
4	95 (19/20)	100 (20/20)	95 (19/20)
5	90 (18/20)	90 (18/20)	90 (18/20)
6	100 (20/20)	95 (19/20)	90 (18/20)
7	95 (19/20)	95 (19/20)	90 (18/20)
8	95 (19/20)	85 (17/20)	95 (19/20)
Total	96 (115/120)	94 (113/120)	93 (111/120)

e.g. Errors: particle → practical; athlete → athlet; rhythm → rthym.

must be grossly intact. Indeed, an argument that has been used against the claim that LBL reading is a perceptual problem is the finding that LBL readers do not show deficits in processing nonlinguistic stimuli, such as objects, pictures of objects, or complex scenes (Patterson & Kay, 1982; Warrington & Shallice, 1980; but see Friedman & Alexander, 1984). And, in fact, many of these patients do not report subjective difficulties in these tasks the way the do for reading. To see whether MJ was similarly unimpaired in processing visual objects, we asked her to perform a series of confrontation naming tasks using a subset of the Snodgrass and Vanderwart set of pictures. On different occasions she was asked to name pictures of objects presented in free vision, at 250msec exposure, and at 100msec exposure (see Table 8). Time-limited presentations were shown centred on a computer screen.

Table 8. Percentage (and No.) Correct Responses in Confrontation Naming at Different Exposure Durations

	MJ	JG	TG
Unlimited exposure	97 (126/130)	96 (250/260)	NA
250msec (centred)	92 (119/129)	NA	NA
100msec (centred)	84 (119/142)	91 (129/142)	100 (142/142)
Examples of errors			
Unlimited	fly → bee	fly → spider	
250msec	fork → spoon		
100msec	beetle → fly, or ant	needle → paintbrush	

In the unlimited exposure duration condition, MJ's performance was almost flawless and not different from JG's performance[4] [MJ vs. JG: $\chi^2(1) = .003$, $P < .9$, n.s]. At highly reduced exposures (100msec), MJ's performance decreased to 84% accuracy, which is significantly different from the performance of the hemianopic control [MJ vs. JG: $\chi^2(1) = 10$, $P < .01$] and the normal control [MJ vs. TG: $\chi^2(1) = 23$, $P < .001$]. Although such performance is indicative of a mild perceptual deficit, MJ's ability to identify pictures of objects remains rather accurate. Furthermore, her errors reflected difficulty in perceiving details of the pictures rather than gross confusion about the identity of the target items. For example, she mistook a beetle for a fly, a rabbit for a squirrel, and a flute for a pen. These findings indicate that MJ's processing of complex visual objects is only mildly affected and not significantly different from that of the hemianopic (and under certain conditions the normal) control.

Discussion

In this section we addressed the issue of whether MJ's performance is comparable to other LBL readers reported in the literature on basic reading and lexical access tasks. The major result is that MJ shows a marked word length effect in reading and recognising written words. In free vision tasks, she took on average 2 to 8 seconds to read three- and eight-letter words—abnormally long response times. Similar effects of length were obtained for stimuli presented in horizontal, vertical, and spaced horizontal displays. A similar length effect was observed in a lexical decision task: Consistently slower responses were obtained for words (3 to 6 seconds) and nonwords (6 to 9 seconds) ranging in length from four to eight letters. Reduced exposures of stimuli decreased reading accuracy, with longer words affected more than shorter ones (93% vs. 86% correct for four- vs. eight-letter words with unlimited exposure). Taken to-

---

[4] JG was administered a larger set of pictures as part of another study. All pictures shown to MJ were included in JG's set.

gether, these results indicate that MJ's reading difficulties are characterised by an abnormal increase in recognition latencies as words get longer, and increasing error rates for longer words at short exposure durations.

MJ's excellent performances in the recognition of orally spelled words, and in oral and written spelling production, indicates that MJ's reading difficulty is limited to the processing of orthographic strings from the visual modality and does not extend to output processes or to "orthographic" processing through the auditory modality. Furthermore, as reported in the Case Histories, MJ demonstrates essentially perfect performance in processing single letters, even when these are displayed at exposures as short as 16msec. And, as indicated by her performance in the visual object naming task, MJ does not have gross perceptual impairments. These results taken together allow the classification of MJ as a classic LBL reader—that is, she has intact lexical orthographic knowledge, she has intact knowledge of graphemes, she does not have gross perceptual impairments, but she has severe difficulties in the rapid recognition of written words.

Although MJ's and JG's reading performances show several similarities, such as a gradual increase in the percentage of errors for letter positions progressively further away from the beginning of the word, JG's reading performance lacks the typical length effect demonstrated by LBL readers. Thus, we can conclude that even though MJ and JG both present with a right visual field cut and a reading deficit characterised by increasing error rates for progressively rightmost letter positions, only MJ shows the characteristic length effect of a LBL reader. In other words, a visual field cut (found in both MJ and JG) is not a *sufficient* condition for being a LBL reader.

MJ's LETTER-BY-LETTER READING IS NOT A LANGUAGE SPECIFIC DEFICIT

This section of our investigation addresses the hypothesis that MJ's LBL reading results from a deficit at a prelexical perceptual stage. One argument that has been used against this hypothesis in the case of other LBL readers is that they do not demonstrate difficulties processing nonlinguistic materials such as pictures. In the previous section we showed that, under certain conditions, this is also the case for MJ. However, more stringent tests are needed before we can rule out the possibility that LBL reading is not the result of a perceptual deficit.

In the following series of tests we assess MJ's ability to process multiple items presented simultaneously in her visual field. Following Warrington and Shallice (1980), we first used a visual span task requiring subjects to detect and report two simultaneously displayed items. Then, following Rapp and Caramazza (1991), we used a letter detection and a line orientation detection task to assess MJ's ability to detect a target item among distracters presented at several spatial locations.

Visual Span

Among other arguments (see the General Discussion), the finding of preserved visual span in some LBL readers led Warrington and Shallice (1980); see also Warrington & Langdon, 1993) to conclude that LBL reading is not the result of a low-level visual processing deficit but rather an inability to process multiple items simultaneously at higher levels of lexical processing.

To assess MJ's and JG's visual span, we used a task originally used by Warrington and Shallice (1980, Experiment 1b). As described in that paper, two numbers were placed on either end of a four-, five-, six-, seven-, and eight-letter word (e.g. 4jury6). Stimuli were printed in lower-case letters, Geneva font size 20, and displayed using the software Psycholab v.1. Items were displayed for 150msec to prevent eye movements. Different word lengths were tested in a fixed order from short to long. There were 5 stimuli in each length condition for a total of 25 stimuli. Subjects were instructed to report the digits. There were two versions of the experiment: one in which stimuli were displayed centred and one in which stimuli were displayed left of fixation. Both versions were administered to MJ, JG, and TG. Results are shown in Table 9.

For the centred condition, both MJ and JG performed significantly more poorly than the normal control in reporting both digits correctly [MJ vs. TG: $\chi^2(1) = 23, P < .001$; JG vs. TG: $\chi^2(1) = 25, P < .001$]. When correct report for the left and the right side digit are considered independently, it can be seen that both MJ and JG reported the digit on the left of the word significantly more often than the digit on the right [MJ_L vs. MJ_R: $\chi^2(1) = 23, P < .001$; JG_L vs. JG_R: $\chi^2(1) = 22, P < .001$]. Distance from fixation point (i.e. intervening word length) did not have any effect on performance for any of the subjects, but this may be an artefact due to the small sample size.

For the left of fixation condition, MJ still had difficulty reporting both digits relative to the normal control [TG vs. MJ $\chi^2(1) = 20, P < .001$]

Table 9. MJ's, JG's, and TG's Performance on the Visual Span Task (% and No. Correct). For Each Display Type (Centred and Left of Fixation), Results Are First Reported for Both Digits, and Then for the Left and the Right Digit Independently

	MJ^a	JG^a	TG^a
Centred stimuli			
Both digits correct	8 (2/25)	0 (0/25)	100 (25/25)
Left digit	100 (25/25)	88 (22/25)	100 (25/25)
Right digit	8 (2/25)	0 (0/25)	100 (25/25)
Left of fixation			
Both digits correct	20 (5/25)	63 (15/25)	100 (25/25)
Left digit	88 (24/25)	96 (24/25)	100 (25/25)
Right digit	20 (5/25)	64 (16/25)	100 (25/25)

[a]No effect of word length; chance level = 10%.

and so did JG [TG vs. JG: $\chi^2(1) = 10$, $P < .01$]. However, when correct report is considered separately for the left-side and the right-side digit, it can be seen that MJ's performance for the right digit improves very little relative to her performance in the centred condition [MJ$_{RC}$ vs. MJ$_{RL}$: $\chi^2(1) = 3$, $P < .1$, n.s.] whereas JG's performance improves considerably [JG$_{RC}$ vs. JG$_{RL}$: $\chi^2(1) = 16$, $P < .001$]. Importantly, distance from the fixation point (i.e. intervening word length) did not have an effect on performance for either subject but, once again, this result must be interpreted cautiously because of the small number of observations.

These results indicate that unlike Warrington and Shallice's (1980) patient RAV, MJ shows a severely reduced visual span even when stimuli are presented in the intact left visual field. The results also show that MJ's reduced span is not simply due to her field cut since she performed considerably worse than the hemianopic control. Thus, MJ's right homonymous hemianopsia cannot entirely account for her LBL reading.

Letter Detection

Following Rapp and Caramazza, (1991, Experiment 3), we used a letter search task to evaluate whether MJ's and JG's ability to detect a target letter among distractors was affected by (1) the absolute spatial location of the target (i.e. whether a target item is presented left, right, or at fixation), and (2) the relative spatial location of the target (i.e. whether a target item is presented in the first, middle, or final position in a string of items). Five- and three-letter strings were presented. There were three versions of the experiment, defined by the position of the letter strings relative to fixation. In one version, the five-letter strings appeared centred across fixation; in another version, the stimuli were displayed with the last letter at the point of fixation (left of fixation condition); and, in another version, the first letter was shown at fixation (right of fixation condition). In the three-letter string conditions, stimuli were presented across the five positions defined by the five-letter string (see Table 10 and Figure 5). Thus, they could either be centred across the five-letter strings' midpoint or aligned with its left or right margin. This allows us (1) to examine performance at different absolute locations relative to fixation (+ 4, + 3, + 2, + 1, F, – 1, – 2, – 3, – 4) while maintaining relative positions constant, and (2) to determine whether performance at any of these positions is affected by the relative position of an item within a string. Thus, for example, with three-letter strings we could examine whether performance at location + 2 differs according to whether the item at that location is the rightmost or leftmost in a string.

The letters K, X, Z, N, and Y were used as target items. The distracters were D, J, G, R, B, and P. There were a total of 280 trials in each version of the experiment. A target was present among distracters on half of the trials (total $N = 140$). Half of the stimuli contained five letters and half three letters, randomly intermixed. The maximum eccentricity of a stimulus was approximately 2.5°. Stimuli were displayed on a black-and-white computer screen using the software Psycholab v.1.

Letters were all capitals, Geneva font size 24, and appeared in black on a white background. All three versions were presented to MJ in a randomised order (total $N = 840$ trials); only the centred version was presented to JG and TG (total $N = 240$ trials).

On each trial a target letter appeared at fixation for 150msec, followed by an ISI of 50msec, and then a string of either three or five letters, which remained on the screen for 150msec (see Table 10 and Fig. 5). Subjects were instructed to keep their eyes at fixation and to press a computer key each time they detected the target item in the letter string.

The results for all three subjects are shown in Table 10. A first comparison was carried out between MJ's overall performance on the centred condition and JG's and TG's performance on the same condition separately for the five- and three-letter string conditions. MJ performed worse than the normal control both in the five-letter [MJ$_5$ vs. TG$_5$: $\chi^2(1) = 23.04$, $P < .001$] and the three-letter strings conditions [MJ$_3$ vs. TG$_3$: $\chi^2(1) = 34.8$, $P < .001$], but she did not perform worse than the hemianopic control in either the five-letter [MJ$_5$ vs. JG$_5$: $\chi^2(1) = 1.96$, $P < .2$] or the three-letter string condition [MJ$_3$ vs. JG$_5$: $\chi^2(1) = 0.64$, $P < .5$].

A second set of analyses was performed to compare MJ's performance on absolute positions at centre, left of fixation, and right of fixation, and on relative initial, middle, and final string positions for three-letter strings. Two-way ANOVAs for three-letter strings

Table 10. Letter Detection Task—% Correct per Display Position

Display type	MJ 1	2	3	4	5	JG 1	2	3	4	5	TG 1	2	3	4	5
Left:					F										
5-letter string	60	30	10	50	60										
3-letter string	100	100	50												
			90	100	80										
				100	100	70									
Centre:			F					F					F		
5-letter string	90	90	70	0	0	100	100	80	20	20	90	100	100	100	100
3-letter string	100	80	50			100	100	80			100	100	100		
			90	90	10		100	100	0			100	100	100	
				90	20	0		100	20	0			100	100	100
Right:	F														
5-letter string	80	10	0	0	0										
3-letter string	90	0	0												
			60	0	0										
				10	0	0									

$N = 10$ per cell; chance level = 10%

Example of Events:		Exposure Duration:
time 0	READY?	
time 1		
time 2	X	150msec
time 3		50msec
time 4	B D X	150msec
response	"yes" if there is a match	
time 6	READY?	

Example of display positions (on each trial either a five- or three-letter single string would appear preceded by a target item):

```
Condition:              Fixation Point
Centred                        +
5-letter strings        1  2  3  4  5
3-letter string         1  2  3
                           1  2  3
                              1  2  3

Left of fixation               +
5-letter strings   1  2  3  4  5
3-letter strings   1  2  3
                      1  2  3
                         1  2  3

Right of fixation              +
5-letter strings               1  2  3  4  5
3-letter strings               1  2  3
                                  1  2  3
                                     1  2  3
```

Fig. 5. Design of Letter Detection Task.

indicate the presence of a main effect of display type [$F(2,18) = 24.65$, $P < .0001$], a main effect of relative string position [$F(2,18) = 13.58$, $P < .0003$], but no interaction [$F(4,18) = 1.99$, $P < 1$, n.s.]. These results indicate that MJ's performance is affected by both absolute (left of vs. right of fixation) and relative (first vs. last letter in a string) position of the items in visual space.

The finding of an effect of absolute letter position is not surprising given MJ's visual field cut. However, there is also an effect of relative position with accuracy systematically lowest at the rightmost position in each three-letter stimulus. This effect occurs for strings of letters presented entirely within the intact left visual field. This can be seen best by comparing performance at the relative position 3 of stimuli presented leftmost in the left display (50%) with performance at the relative position 1 of stimuli presented rightmost in the left display (100%). These results are consistent with those obtained in the visual span task, where it was found that MJ could report the left digit correctly but made many errors for the right digit.

Line Orientation Detection

In order to investigate whether MJ's spatially determined deficit extends to the processing of nonlinguistic stimuli, she was tested on an orientation detection search task identical in design to the letter search task discussed earlier. The task was designed to evaluate whether MJ's ability to detect a target item among distracters was affected by the absolute and/or relative spatial location of the target. Five- and three-item strings were presented.

Horizontal (–), vertical (|), and tilted to the right (/) lines were used as target items. The same items were used as distracters, but on each trial target items never occurred twice (e.g. target: "–"; five-item string " | / – | / "). There were a total of 168 trials in each version of the experiment. A target was present among distracters on half of the trials (total $N = 84$). Half of the stimuli contained five lines and half three lines, randomly intermixed. The maximum eccentricity was approximately 2.5°.

Stimuli were displayed on a black-and-white computer screen using the software Psycholab v.1. Lines were displayed in bold, size 24, and appeared in black on a white background. All three versions of the experiment were given to all subjects, with stimuli within and across conditions appearing in a randomised order (total N = 504 trials).

On each trial a target line appeared at fixation for 150msec, followed by an ISI of 50msec, and then a string of either 3 or 5 lines, which remained on the screen for 150msec. Subjects were instructed to keep their eyes at fixation and to press a computer key each time they detected the target item in the string (see Fig. 5).

MJ's and JG's performance is shown in Table 11a, and TG's performance is shown in Table 11b. MJ's overall performance, collapsing across left, centred, and right conditions was compared to JG's and TG's overall performance on the same conditions. MJ performed less accurately than the normal control for both five- [MJ$_5$ vs. TG$_5$: $\chi^2(1)$ = 21.16, P < .001] and three-item strings [MJ$_3$ vs. TG$_3$: $\chi^2(1)$ = 16.81, P < .001], but did not differ from the hemianopic control for either five- [MJ$_5$ vs. JG$_5$: $\chi^2(1)$ = 0.09, P < .8] or three-item strings [MJ$_3$ vs. JG$_3$: $\chi^2(1)$ = 2.01, P < .95].

A second analysis was performed to compare MJ's performance on absolute positions (left vs. right of fixation), and on relative initial, middle, and final string positions for three-item strings. A two-way ANOVA indicated the presence of a main effect of fixation [$F(1,12)$ = 19, P < .00], a main effect of relative string position [$F(2,12)$ = 5.9, P < .02], but no interaction [$F(2,12)$ = .9, P < .4]. These results replicate the previous findings in the letter detection task, and indicate that MJ's performance for nonlinguistic items is affected by both absolute (left vs. right of fixation) and relative (first vs. last item in a string) position of the items in visual space.

The same type of analysis for JG's results revealed a main effect of fixation [$F(1,12)$ = 64, P < .001], but no effect of relative string position [$F(2,12)$ = 2.7, P < .1, n.s.], nor of interaction [$F(2,12)$ = 1.6, P < .24, n.s.].

Discussion

Taken together, the findings reported in this section indicate that (1) MJ presents with a deficit that is not entirely accounted for by her right visual field cut, as errors occur also for items displayed entirely within the good visual field; (2) this deficit affects performance for linguistic and nonlinguistic items alike; and (3) performance is characterised by increasing difficulty in processing progressively more right parts of the stimuli.

These results are consistent with previous reports (e.g. Kay & Hanley, 1991; Montant, Nazir, & Poncet, this issue), which indicate that processing of arrays of items (linguistic and nonlinguistic) is deficient in some LBL patients, and that this may be the source of the observed LBL reading performance.

The results obtained thus far also indicate that MJ's deficit is dissimilar from other cases of LBL readers in various respects. As already noted, MJ's deficit involves a reduction of vis-

Table 11a. Orientation Detection Task: MJ's and JG's Performance (% Correct) per Display Position

	MJ					JG				
Display Type	1	2	3	4	5	1	2	3	4	5
Left:					F					F
5-item	83	83	33	33	50	67	100	83	100	100
3-item	83	83	67			100	100	100		
		100	83	67			100	100	83	
			75	67	42			100	83	100
Centre:			F					F		
5-item	83	83	83	50	17	83	100	83	0	0
3-item	83	67	42			100	83	100		
	100	100	56				100	83	0	
			83	58	8			100	33	0
Right:	F					F				
5-item	100	67	17	0	0	100	17	0	0	0
3-item	83	58	8			100	0	0		
		83	17	0			33	0	0	
			17	0	0			0	0	0

$N = 6$ per cell; chance = 17%.

ual span, a finding that contrasts with other reports in the literature (e.g. case RAV of Warrington & Shallice, 1980). Also, MJ shows nor-

Table 11b. Orientation Detection Task: TG's Performance (% Correct) Per Display Position

	Relative Bar Position				
	1	2	3	4	5
All displays[a]					
5-item	100	100	89	100	100
3-item	100	100	100		
		100	100	100	
			89	100	100

$N = 18$ per cell.
[a]Left, centre, and right displays collapsed; *No effect on RT.

mal, rapid processing of single letters in contrast with numerous reports in the literature (Arguin & Bub, 1994; Bub & Arguin, 1995; Rapp & Caramazza, 1991). However, MJ's performance is also similar to that of other patients reported in the literature. For example, she shows mild difficulties in the processing of pictures of objects (Friedman & Alexander, 1984), and presents with a right homonymous hemianopsia. Some of the comparisons may not be appropriate, however. For example, case RAV (Warrington & Shallice, 1980) does not present with several characteristics that are commonly found in other LBL readers, and other LBL readers reported in the literature have not typically been tested for visual span.

Thus, we do not know whether a reduced visual span is or is not commonly associated with LBL reading. Nonetheless, the fact that visual span is not necessarily associated with the LBL reading disorder indicates that there may be different causes for this complex disorder. Similarly, since single-letter processing difficulties are not necessarily associated with the LBL reading disorder we may infer that different forms of perceptual deficits may contribute to the disorder.

The results presented thus far have served to demonstrate that MJ presents with the characteristic performance observed in LBL readers, and that, unlike some other cases, her deficit does not affect letter processing but is characterised by a reduction of visual span. Given these input processing characteristics, how might visual word recognition occur? Are the input conditions adequate for lexical access in the absence of explicit identification of words, as has been reported for other LBL readers (Buxbaum & Coslett, 1997; Coslett & Saffran, 1989; Coslett & Saffran, 1994a, 1994b). We address these questions in the following section.

DOES MJ SHOW IMPLICIT READING?

In order to address the question of whether lexical access can be achieved even when altered input processing may not be sufficient for recognition, we first tried to replicate previous findings reported in the literature (Coslett & Saffran, 1989; Coslett et al., 1993; Shallice & Saffran, 1986; Warrington, 1975; Warrington & McCarthy, 1983). It has been shown that some dyslexic patients' performance in categorisation tasks can be significantly above chance level even when items are presented at exposures too brief for overt reading and recognition. For example, Warrington and Shallice (1979) report the case of AR who, in a forced-choice task, could decide the semantic category of a word despite an inability to read it. We first tried to replicate this effect in a recognition task, and then we manipulated chance-level performance and visual similarity features between target and response in a word–picture matching task to see how these factors affect categorisation performance.

Semantic Categorisation

Following previous investigations of patients with alexia (Rapp & Caramazza, 1989; Warrington & Shallice, 1979) MJ was given a word categorisation task. In this task she was asked to read words individually displayed on a computer screen, and for those words that she failed to read she was asked to make a forced-choice decision about the semantic category to which the word belonged.

MJ was presented with a list of 161 words. The words were divided into 3 lists consisting of 54, 59, and 48 words, respectively. Within each list there were words belonging to three different categories (matched for frequency, letter length, and syllable length), such that the same three category names were presented for each categorisation judgement. For example, list 1 consisted of names of places, means of transportation, and household items.

Words were displayed centred for 200msec[5]—well below MJ's threshold for accurate recognition—and she was instructed to read aloud each word, guessing if unsure. If she read a word correctly, the trial was discarded, as were cases in which she produced an incorrect response. For cases in which she produced a "don't know" response (even after encouragement to guess), she was asked to choose from among three aurally presented categories the one to which she thought the word might belong. For example, if in list 1 MJ was presented with the word "truck", and if she was unable to give a lexical response, she was asked whether the word was a place, a means of transportation, or a household item.

The results are reported in Table 12, and show that 42% (67/161) of the words were read correctly, 23% (37/161) were read incorrectly, and 35% (57/161) elicited a "don't know" response. Categorisation performance with the 57 "don't know" items indicates a 60% accuracy level (34/57), which is significantly above chance [$\chi^2(1) = 7.29$, $P < .01$]. These results "replicate" the finding that LBL readers may be able to access the meaning of words they are unable to pronounce.

Word–Picture Matching

It has been argued that patients' above-chance performance on categorisation tasks, like the

Table 12. Categorisation Task—% of Responses per Category

	Reading			"Don't knows" Correct Categorised
	Correct	Incorrect	"Don't knows"	
List 1				
Places	68	11	21	100*
Transport	53	13	34	100*
Household	35	20	45	60
List 2				
Animals	45	20	35	43
Foods	79	11	10	100*
Body parts	5	50	45	33
List 3				
Clothing	36	14	50	43
Names	26	26	48	60*
Professions	27	40	33	40
Total	42	23	35	60*

Exposure: 200msec; presentation: centred; categorisation chance = 33%.
*Significant $P < .01$.

one just described, may reflect direct access to superordinate semantic information despite failure to access the word's specific meaning (Warrington & Shallice, 1979). Findings of this sort have also been used to support the hypothesis of a right-hemisphere lexical/semantic system that in such patients is disconnected from the left-hemisphere output lexical/phonological system (see General Discussion). That is, MJ's performance in the

---

[5] In this and in the following experiments, exposure duration was set at the fastest speed that MJ could tolerate. When exposures below 200msec were attempted she reported that she could not see anything but a flash on the screen, and when encouraged to guess what the word might have been she stated that she was unable to do so.

categorisation task may be interpreted as indicating access to the right-hemisphere semantic system. For example, when presented with the word "TRAIN" she would be able to access the information that this is a "means of transportation" even though she would be unable to identify the specific sequence of letters (or sounds) that distinguish, say, "TRAIN" from "TRAM" or "TRAY". On this interpretation, items sharing a superordinate (i.e. semantic class) would be indistinguishable in terms of meaning (i.e. MJ would not be able to distinguish between "TRAIN" and "TRAM", but would be able to distinguish between "TRAIN" and "TRAY"). Thus, this hypothesis predicts that items belonging to the same category ("TRAIN" and "TRAM") should be confused more often than items belonging to different categories ("TRAIN" and "TRAY"), since in the latter case knowledge of their category membership should be sufficient to allow the distinction.

To test this prediction a word–picture matching task was designed in which two variables were manipulated—the degree of semantic and orthographic similarity between each word and the name of the picture presented for matching (Rapp & Caramazza, 1989). In this task, MJ was first presented with words displayed for 200msec and was asked to read them aloud. To ensure maximum processing of the visual input, items were displayed left of fixation. As before, items that she read either correctly or incorrectly were discarded. If she produced a "don't know" response, she was then presented with a picture and asked to decide whether the word and the picture matched. On 30% ($N = 170$) of the trials the word and the picture constituted a match (e.g. the word "TOOTH" and the picture of a tooth). The other 70% ($N = 395$) of the trials consisted of 6 categories, defined by different kinds of relationship between the word and the picture. Word and picture were either semantically similar (SS) (i.e. belonged to the same superordinate category—e.g., the word "SOCK" and the picture <glove>[6] from the category clothing) or semantically dissimilar (SD) (e.g. PLANE—<mouth>); and they were either visually similar (VS) (e.g. TAXI—<tank>) or visually dissimilar (e.g. PILOT—<dentist>). Two levels of visual similarity were used: for level A (low similarity), the words in the pair shared at least the first letter and had approximately the same length (e.g. PEAR—<plum>); for level B (high similarity), the word and picture name shared the first letter, they were of approximately the same length, and they shared at least 50% of the letters (e.g. TOE—<tie>). Visual dissimilarity (VD) is defined by the absence of a shared initial letter, and by fewer than 50% letters in common (e.g. MOUTH—<suit>). Based on these criteria, the following six categories were defined: (1) VD, SS: word and picture are visually dissimilar but semantically similar (e.g. SOCK—<glove>); (2) VD, SD: word and picture are visually and semantically dissimilar (e.g. PIE—<bow>); (3) VSA, SS: word and picture are visually similar at level A and share a

---

[6] With the symbolic representation < . . . > we mean the picture of the designated item.

superordinate category (e.g. PEAR—<plum>); (4) VSA, SD: word and picture are visually similar at level A but do not share a superordinate category (e.g. DOMINO—<donkey>); (5) VSB, SS: word and picture are visually similar at level B and share a superordinate category (e.g. TAXI—<tank>); (6) VSB, SD: word and picture are visually similar at level B but do not share a superordinate category (e.g. TOE—<tie>). Word lists for each of the conditions were approximately matched for average frequency. Test items were presented in five blocks, which were administered in successive testing sessions, with at least one non-testing day between each session.

MJ's performance on this task is summarised in Table 13, which indicates that 49% (276/561) of the words were read correctly, 16% (88/561) were read incorrectly, and 35% (197/561) elicited a "don't know" response. There were four missed responses (when MJ got distracted while the word was flashed on the screen). Matching performance with the 197 "don't know" trials led to 84.2% (48/57) correct "yes" responses, and 74.3% (104/140) correct "no" responses. This performance is significantly different from chance level [$\chi^2(1)$ = 57.1, $P < .001$]. In particular, she accepts correct matches (e.g. TOOTH—<tooth>) at a rate significantly *above* chance [$\chi^2(1) = 367$, $P < .001$] and rejects incorrect matches at a rate significantly *above* chance [$\chi^2(1) = 67, P < .001$]. More importantly though, when matching performance for the "no" responses is analysed in terms of the semantic similarity factor, MJ's performance appears *not* to be affected by the semantic similarity between word and picture. Thus, she is just as likely to reject the

Table 13. Word–picture matching task

		Reading Response (%)			Categorisation Response (%)	
Pair Type	Example	Correct	Incorrect	DK	Yes	No
Match	tooth–tooth	52	15	33	84	16
VD,SS	jaguar–tiger	57	17	26	0	100
VD,SD	plane–mouth	52	15	33	0	100
VSA[a],SS	butter–bun	47	17	36	8	92
VSA,SD	glove–ghost	35	12	53	22	78
VSB[b]–SS	taxi–tank	49	16	35	69	31
VSB,SD	pipe–pine	49	17	34	65	35
Total		49	16	35	77	23

VD = Visually Different; VS = Visual Similar; SD = Semantically Different; SS = Semantically Similar.
[a]Type A indicates low visual similarity.
[b]Type B indicates high visual similarity.
Exposure: 200msec; presentation: left of fixation; chance level: 50%.

this version of the experiment, 23 responses were eliminated for MJ and 6 responses for TG. MJ's and TG's performance for the different categories is reported in Table 14.

As in the picture version of the experiment, responses were analysed after collapsing across hand and target category (animal and artefact). The results indicate that TG again shows a significant effect of identity priming relative to the pattern condition ($t = 2.2, P < .03$), the unrepeated condition ($t = 2.35, P < .02$), and the mixed condition ($t = 2.75, P < .007$). MJ did not show any facilitatory effects.

These results indicate that in spite of the fact that MJ demonstrates identity priming effects for pairs of repeated pictures, she shows no facilitation in her recognition of a picture with previous exposure to the name of the picture. Thus, there is no evidence that lexical and semantic activation have occurred in MJ's lexical system upon exposure to the prime word. In other words, we find no evidence of implicit reading in MJ.

Discussion

In this section we addressed the issue of whether implicit reading occurs in MJ. Using a categorisation task, we first replicated findings reported in the literature which indicate that "implicit reading" occurs for words that cannot be identified overtly. That is, MJ showed above-chance performance in categorisation of words that she failed to read. However, we then went on to show that the above-chance performance displayed by MJ is best accounted for in terms of sophisticated guessing. This conclusion is supported by findings obtained in a word–picture matching task, where MJ was asked either to accept or reject a picture as an appropriate match to a word that she had failed to read. In this experiment, we manipulated the degree of visual similarity between the input word and the name of the pictorial foil. The results indicate that at sufficient levels of visual similarity. MJ's performance deteriorates significantly below chance level as she systematically accepts the foils as correct matches. That is, when the information that she obtains from the initial part of the word is not sufficient to discriminate between a correct match and a foil, she appears not to have any further information available to help her constrain her responses. In other words, MJ does not appear to have access to lexical information for those word stimuli she has difficulty processing at short exposure durations. This is true for stimuli displayed entirely in the intact visual field. On this interpretation, MJ has access to words only after a laborious LBL analysis of the input stimulus, and her ability to categorise words she is unable to read is due to sophisticated guessing. That is, MJ's ability to extract *some* visual information from briefly presented words allows her to generate potential lexical candidates for response which, when paired with knowledge that the response is a member of a small set of categories, may be used to direct her choice to the correct category or item.

It has been claimed that implicit reading may only be expected under conditions in which the patient is discouraged from using an LBL reading strategy and is encouraged instead to use a "whole-word" recognition approach (Coslett & Saffran, 1994a, 1994b). Thus, it could be argued that MJ's performance in the categorisation and the word–picture matching tasks in which she was first asked to read the word may have reduced the likelihood of her using a whole-word reading strategy. However, this objection does not apply to the "identity" priming experiment. In this experiment, MJ was not required to read the priming words, but simply to perform a categorisation (animal vs. artefact) of the target picture. Both MJ and the normal control, TG, showed facilitatory effects of identity priming between pictures. However, unlike TG, MJ does not show such facilitation between a target picture and a prime word, even when both are displayed left of fixation. These results indicate that only visual iconic stimuli can be processed accurately and quickly in the right hemisphere and lexical stimuli cannot. In short, the results we have reported fail to demonstrate "implicit reading" in MJ and support the conclusion that (at least in her case) the priming effects observed in the categorisation and word–picture matching tasks merely reflect sophisticated guessing.

GENERAL DISCUSSION AND CONCLUSIONS

We have reported the case of MJ, who suffered a CVA that involved the left optic radiation near the thalamus and the corpus callosum, and that resulted in a complete right homonymous hemianopsia and a reading deficit characterised by a laborious letter-by-letter processing. However, she showed normal language abilities and preserved copying, writing, oral spelling, letter naming, and aural recognition of spelled words.

Four features of MJ's performance are of particular relevance here: (1) MJ has no difficulty processing single letters, as indicated by her normal letter naming performance with stimuli displayed for very brief exposure durations (16msec); (2) despite (1), she has serious difficulties in processing multiple objects simultaneously, as indicated by her performance in the visual span task and the letter and line segment detection tasks, which showed a processing deficit characterised by a left-to-right gradient of difficulty for both absolute and relative positions in visual space; (3) she appears to have no difficulty accessing the orthographic input lexicon when presented with letter names for aural recognition, or when allowed enough time to process a visually presented letter string; and (4) despite (1) and (3), she does not have normal access to lexical orthographic representations and their meanings when stimuli are presented for too brief a time to allow for serial processing of the letter string. Indeed, we found no evidence of (partial) lexical or semantic access without corresponding recognition of the letters in a word: No signs of implicit reading were observed when the input stimuli were controlled for the relevant visual features; "implicit reading" was only obtained under conditions that allow sophisticated guessing.

How can MJ's pattern of performance be understood in terms of previous interpretations of LBL reading? Four broad classes of explanations have been offered for the LBL reading disorder: (1) a deficit at the level of the orthographic lexicon, (2) a deficit in accessing intact lexical orthographic representations, (3) a letter processing deficit, and (4) a basic perceptual deficit. We will briefly address each class of explanation in light of the results obtained with MJ.

Warrington and her collaborators (Warrington & Langdon, 1994; Warrington & Shallice, 1980) originally offered the interpretation that LBL reading is the result of damage to the orthographic lexicon (which the authors call the "visual word form system"). Damage to this system led patients to the strategic use of the spelling system as an alternative mode of word recognition. The adoption of this processing strategy results in a complete by-passing of the damaged orthographic lexicon in "reading." Presumably, LBL readers feed the individual letter names of a written word into this spelling system, which then functions in reverse to recognise the letter string as a word. Thus, as long as the spelling system is intact, reverse spelling would allow LBL reading to occur. The results we have reported for MJ are not inconsistent with this account of LBL reading: MJ named letters normally and she could spell perfectly well. The problem with this interpretation of LBL reading is not in fitting the facts but with the account itself. It is entirely mysterious what "reverse spelling" refers to: It is unclear how a processing system, which in the undamaged brain activates a string of graphemes in parallel from a lexical representation (see Caramazza & Hillis, 1990), would, in conditions of brain damage, function in reverse order and serially. Until the "reverse spelling" hypothesis is articulated in greater detail and independent justification is given for its presumed "reverse" functioning, we will have to withhold judgement on its explanatory value for LBL reading.

Patterson and Kay (1982; see also Bowers et al. 199b; Bub & Arguin, 1995; Feinberg, Dyckes-Berke, Miner, & Roane, 1995; Katz, 1980; Mimura, Goodglass, & Milberg, 1996) have argued that LBL reading results from (an unspecified) deficit in accessing an otherwise intact orthographic input lexicon. The authors maintain that orthographic context effects, such as those underlying the word superiority effect, and the observation of semantic priming indicate that LBL readers are able to gain lexical access even under conditions of impoverished input that does not allow for full recognition. This evidence is taken as an indicator of preserved lexical orthographic structure and processing. We agree with these authors that findings of this sort indicate preserved lexical orthographic knowledge. We also agree that such knowledge could be accessed through partial activation of lexical entries following presentation of an orthographic input that cannot be recognised overtly by the patient under limited exposure duration. However, we found no evidence in word-to-picture priming experiments that MJ was able to access semantic information unless she recognised the word (see Schacter, Rapcsak, Rubens, Tharan, & Laguna, 1990). In her case, above-chance performance in semantic categorisation tasks was most likely to be the result of sophisticated guessing. Whether a similar claim can be made for the cases cited in support of the lexical access deficit hypothesis of LBL reading remains an open question. And, in any case, this account leaves unexplained the relative letter position effects observed with MJ (see also Montant, Nazir, & Poncet, this issue).

Various authors have shown that some LBL readers present with a deficit at the level of letter processing (Arguin & Bub, 1994; Behrmann & Shallice, 1995; Howard, 1990; Rapp & Caramazza, 1991). Some have concluded that this deficit is causally linked to a patient's LBL reading (Arguin & Bub, 1993, 1994; Behrmann & Shallice, 1995; Montant et al., this issue), while others have viewed the letter processing deficit as functionally independent from it (Bowers et al., 1996a; Hanley & Kay, 1996; Kay & Hanley, 1991; Rapp & Caramazza, 1991). In the present investigation, we have taken MJ's perfect performance in letter naming at 100msec exposure and nearly perfect at 16msec exposure as an indication of preserved *single* letter processing. We agree with Behrmann and Shallice (1995, p. 452) who conclude that "... even in those cases in whom individual letter recognition is considered to be accurate, identification may well be slowed" and that more stringent assessment is needed before concluding that processing is normal. However, also in agreement with these authors, we believe that an account that interprets LBL reading as resulting from "slowed letter processing" is compatible with the interpretation that the reading disorder results from a more basic visual perceptual deficit that is not specific to letter processing. In support of the latter interpretation, we note that MJ showed comparable processing deficits in processing letter strings and simple line segments in time-constrained visual search tasks.

Finally, there is a class of interpretations that considers LBL reading to be the result of a basic perceptual deficit that is not specific to letter stimuli (Farah & Wallace, 1991; Friedman & Alexander, 1984; Kay & Hanley, 1991; Rapp & Caramazza, 1991). For example, Rapp and Caramazza (1991) reported the case of a LBL reader who, like MJ, showed a left-to-right gradient of processing difficulty for linguistic and nonlinguistic visual inputs alike. These authors suggested that LBL reading may result from a deficit that impairs the even allocation of processing resources across all spatial locations of a visual input. Because of this unequal allocation or deployment of resources, the encoding of information occupying progressively rightmost positions in the stimulus may be severely restricted. Consequently, repeated, serial fixations of letters (or other objects) becomes necessary in order to process an array of items for recognition. This account is consistent with the pattern of performance obtained with MJ.

Although the perceptual deficit explanation of LBL reading is consistent with the results reported for MJ, it fails to address the issue of why a presumably undamaged right hemisphere should show a basic perceptual processing deficit. One possible explanation might be that the right visual processing system does not allocate processing resources evenly across visual space resulting in inadequate processing of the rightmost part of stimulus strings.

However, this explanation is manifestly wrong. MJ showed the same performance for horizontally and vertically displayed words, and we could not induce a significant decrement in her abilty to process pictures of objects either in recognition or priming conditions[10]. Furthermore, the hemianopic control subject did not show a similar reading deficit. Thus, the hypothesis that the right-hemisphere visual processing system is somehow limited in its capacity to process visual strings must be considered with great scepticism.

The explanation that seems most compatible with the results we have reported for MJ is as follows: Visual information is processed normally by the intact right hemisphere but because this hemisphere (at least in some subjects) cannot interpret letters as graphemes it transfers visual form information to the left hemisphere. Since the units of interpretation are letters, this results in a serial scan and transmission of letter form information. (A similar process may occur with line segments if we assume that the response decision mechanism is located in the left hemisphere.) This process is necessarily slow and not compatible with the normal parallel functioning of word recognition in the left hemisphere, resulting in error-prone performance. Depending on the assumptions we make about the nature of the information transferred across hemispheres, and the nature of the degradation of this information as a consequence of the partial disconnection of the hemispheres, other explanations of LBL reading are possible.

The same set of evidence that has been invoked to indicate preserved lexical knowledge and its activation in the left hemisphere has also been invoked to argue for lexical activation and the representation of knowledge in the right hemisphere. In particular, in a series of reports Coslett and Saffran (1994a, 1994b) have argued that the "lexical" effects (semantic priming and good lexical decision performance) they obtained with their pure alexic patients are best understood as reflecting the residual reading capabilities of the right hemisphere (see also Coltheart, 1983; Zaidel & Schweiger, 1984; but see Patterson & Besner, 1984). This is an interesting possibility. However, if we were to agree with Coslett and Saffran, it would be unclear why MJ did not show comparable lexical effects to those observed with other LBL readers (see also Miozzo & Caramazza, this issue). Of course, it can be argued that there is considerable vari-

---

[10] However, there are two possible interpretations for this effect (or lack thereof). First, this could be a true effect, indicating that processing of pictorial information is appropriately processed in the right hemisphere and is not affected by damage to the mechanism responsible for the "LBL reading" deficit. Alternatively, this result may be a by-product of the object recognition task, which does not require exhaustive parsing of the visual information provided by the input stimulus. It may be possible to correctly identify an object on the sole basis of partial and even minimal information (e.g. a handle for a cup, or whiskers for a cat). In other words, it is possible that recognising a word or a string of characters is a more demanding task than recognising a picture of an object, in spite of the qualitative and functional equivalence of the two tasks. If this latter interpretation were true, then under conditions that prevent recognition on the basis of partial information on this task should also show the characteristic gradual decrement on the right. We are not aware of evidence addressing this issue, and further investigations are certainly needed.

ation among the reading capabilities of right hemispheres. And this possibility must be given serious consideration in the light of the results with split-brain patients (Baynes, 1992; Baynes & Eliassen, 1998; Gazzaniga, 1983; Gazzaniga, Smylie, Baynes, McCleary, & Hirst, 1984). The possibility that variation in performance across LBL readers reflects (among other things) different degrees of language specialisation in the patients' two hemispheres highlights the importance of administering the same (or compable) tests across individuals. However, it must also be emphasised that this hypothesis is extremely powerful and not easily testable, since we do not have a way of determining the strength of language lateralisation independently across individuals. Thus, it remains unclear when the failure to show implicit processing (as in the present case) would constitute evidence against the right-hemisphere reading hypothesis.

In conclusion, it would seem that Déjèrine's (1892) interpretation of pure alexia as a deficit reflecting the degraded transfer of information from a normal visual processing system in the right hemisphere to a normal language processing system in the left hemisphere continues to be the most plausible account of this class of disorders. Our findings support the view that LBL reading has (at least in MJ) a prelexical or "perceptual" basis, although there is not sufficient evidence to rule out additional impairments at higher levels of processing as possible sources of the deficit in other cases. The evidence we have reported can be entirely accounted for in terms of the serial transmission of information from the right to the left hemisphere and normal lexical processing in the left hemisphere. This evidence does not speak to the issue of whether there is (at least partial) lexical knowledge in the right hemisphere.

REFERENCES

Arguin, M., & Bub, D. (1993). Single-character processing in a case of pure alexia. *Neuropsychologia, 31*(5), 435–458.

Arguin, M., & Bub, D. (1994). Functional mechanisms in pure alexia: Evidence from letter processing. In M.J. Farah & G. Ratcliff (Eds). *The neuropsychology of high-level vision: Collected tutorial essays* (pp. 149–171). Hillsdale, NJ: Lawrence Erlbaum Associates Inc.

Baynes, K. (1992). Reading with a limited lexicon in the right hemisphere of a callosotomy patient. *Neuropsychologia, 30*(2), 187–200.

Baynes, K., & Eliassen, J.C (1998). The visual lexicon: Its access and organization in commissurotomy patients. In M. Beeman & C. Chiarello (Eds.), *Right hemisphere language comprehension* (pp. 79–104). Mahwah, NJ: Lawrence Erlbaum Associates Inc.

Behrmann, M., & Shallice, T. (1995). Pure alexia: A nonspatial visual disorder affecting letter activation. *Cognitive Neuropsychology, 12*(4), 409–454.

Bowers, J.S., Arguin, M., & Bub, D. (1996a). Fast and specific access to orthographic knowledge in a case of letter-by-letter surface alexia. *Cognitive Neuropsychology, 13*(4), 525–567.

Bowers, J.S., Bub, D., & Arguin, M. (1996b). A characterisation of the word superiority effect in a case of letter-by-letter surface alexia. *Cognitive Neuropsychology, 13*(3), 415–441.

Bub, D.N., & Arguin, M. (1995). Visual word activation in pure alexia. *Brain and Language, 49*(1), 77–103.

Bub, D.N., Black, S., & Howell, J. (1989). Word recognition and orthographic context effects in an letter-by-letter reader. *Brain and Language, 36*, 357–376.

Buxbaum, L.J., & Coslett, H.B. (1997). *Deep dyslexic phenomena in a letter-by-letter reader.* Manuscript in preparation.

Caramazza, A., & Hillis, A. (1990). Levels of representation, co-ordinate frames, and unilateral neglect. *Cognitive Neuropsychology, 7*, 391–445.

Coltheart, M. (1983). The right hemisphere and disorders of reading. In A. Young (Ed.), *Functions of the right cerebral hemisphere* (pp. 171–201). London: Academic Press.

Coslett, H.B., & Saffran, E.M. (1989). Evidence for preserved reading in "pure alexia". *Brain, 112*(2), 327–359.

Coslett, H.B., & Saffran, E.M. (1994). Mechanisms of implicit reading in alexia. In M. Farah & G. Ratcliff (Eds.), *The Neuropsychology of high-level vision: Collected tutorial essays* (pp. 299–330). Hillsdale, NJ: Lawrence Erlbaum Associates Inc.

Coslett, H.B., Saffran, E.M., Greenbaum, S., & Schwartz, H. (1993). Reading in pure alexia: The effect of strategy. *Brain, 116*(21), 21–37.

Damasio, A.R., & Damasio, H. (1986). Hemianopia, hemiachromatopsia and the mechanisms of alexia. *Cortex, 22*, 161–169.

Déjèrine, J. (1892). Contribution a l'etude anatomo-pathologique et clinique des differentes varietes de cecite verbale. *Memoires Societe Biologique, 4*, 61–90.

Farah, M.J., & Wallace, M.A. (1991). Pure alexia as a visual impairment: A reconsideration. *Cognitive Neuropsychology, 8*, 313–334.

Feinberg, D.E., Dyckes-Berke, D., Miner, C.R., & Roane, D.H. (1995). Knowledge, implicit knowledge and metaknowledge in visual agnosia and pure alexia. *Brain, 118*(3), 789–800.

Friedman, R.B., & Alexander, M.P. (1984). Pictures, images, and pure alexia: A case study. *Cognitive Neuropsychology, 1*, 9–23.

Friedman, R.B., & Hadley, J.A. (1992). Letter-by-letter surface alexia. *Cognitive Neuropsychology, 9*(3), 185–208.

Gazzaniga, M.S. (1983). Right hemisphere language following brain bisection: A 20-year perspective. *American Psychologist, 38*, 525–549.

Gazzaniga, M.S., Smylie, C.S., Baynes, K., McCleary, C., & Hirst, W. (1984). Profiles of right hemisphere language and speech following brain bisection. *Brain and Language, 22*, 206–220.

Geschwind, M. (1965). Disconnection syndromes in animal and man. *Brain, 88*, 237–294.

Greenblatt, S.H. (1973). Alexia without agraphia or hemianopia. *Brain, 96*, 307–316.

Hanley, J.R., & Kay, J. (1996). Reading speed in pure alexia. *Neuropsychologia, 34*(12), 1165–1174.

Howard, D. (1990). Letter-by-letter readers: Evidence for parallel processing. In Besner & Humphreys (Eds.), *Basic processes in reading.* Hillsdale, NJ: Lawrence Erlbaum Associates Inc.

Katz, R.B. (1990). Cross-modality word matching in letter-by-letter readers. *Cortex, 26*, 65–76.

Kay, J., & Hanley, R. (1991). Simultaneous form perception and serial letter recognition in a case of letter-by-letter reading. *Cognitive Neuropsychology, 8*, 249–273.

Kinsbourne, M., & Warrington, E.K. (1962). A disorder of simultaneous form perception. *Brain, 85*, 461–486.

Levine, D.H. & Calvanio, R. (1978). A study of the visual defect in verbal alexia-simultanagnosia. *Brain, 101*, 65–81.

Mimura, M., Goodglass, A., & Milberg, W. (1996). Preserved semantic priming effect in alexia. *Brain and Language, 54*(3), 434–446.

Miozzo, M., & Caramazza, A. (this issue). Varieties of pure alexia: The case of failure to access graphemic representations. *Cognitive Neuropsychology, 15*(1/2).

Montant, M., Nazir, T., & Poncet, M. (this issue). Pure alexia and the viewing position effect in printed words. *Cognitive Neuropsychology, 15*(1/2).

Patterson, K., & Besner, D. (1984). Is the right hemisphere literate? *Cognitive Neuropsychology, 1*(4), 315–341.

Patterson, K., & Kay, J. (1982). Letter-by-letter reading: Psychological description of a neurological syndrome. *Quarterly Journal of Experimental Psychology, 34A*, 411–441,

Rapp, B., & Caramazzza, A. (1989). General to specific access to word meaning: A claim re-examined. *Cognitive Neuropsychology, 6*(2), 251–272.

Rapp, B., & Caramazza, A. (1991). Spatially determined deficits in letter and word processing. *Cognitive Neuropsychology, 8*, 275–311.

Reuter-Lorenz, P.A., Brunn, J.L. (1990). A prelexical basis for letter-by-letter reading: A case study. *Cognitive Neuropsychology, 7*(1), 1–20.

Schacter, D.L., Rapscak, S.Z., Rubens, A.B., Tharan, M., & Laguna, J. (1990). Priming effects in a letter-by-letter reader depend upon access to the word form system. *Neuropsychologia, 28*(10), 1079–1094.

Shallice, T., & Saffran, E. (1986). Lexical processing in the absence of explicit word identification: Evidence from a letter-by-letter reader. *Cognitive Neuropsychology, 3*, 429–458.

Warrington, E.K. (1975). The selective impairment of semantic memory. *Quarterly Journal of Experimental Psychology, 27*, 635–657.

Warrington, E.K., & Langdon, D. (1994). Spelling dyslexia: A deficit of the visual word-form. *Journal of Neurology, Neurosurgery & Psychiatry, 57*, 211–216.

Warrington, E.K., & McCarthy, R. (1983). Category-specific access dysphasia. *Brain, 106*, 859–878.

Warrington, E.K., & Shallice, T. (1979). Semantic access dyslexia. *Brain, 102*, 43–63.

Warrington, E.K., & Shallice, T. (1980). Word-form dyslexia. *Brain, 103*, 99–112.

Zaidel, E., & Schweiger, A. (1984). On wrong hypotheses about the right hemisphere: commentary on K. Patterson and D. Besner "Is the right hemisphere literate?" *Cognitive Neuropsychology, 1*(4), 351–364.

VARIETIES OF PURE ALEXIA: THE CASE OF FAILURE TO ACCESS GRAPHEMIC REPRESENTATIONS

Michele Miozzo and Alfonso Caramazza
Harvard University, Cambridge, USA

We document the case of a patient (GV) who, following a left posterior brain lesion, showed a selective and severe deficit in naming visual objects and in reading letters, words, and numerals. Three sets of findings are critical for the interpretation of the patient's alexia. First, despite intact visual processing abilities and preserved ability to recognise the shape and orientation of letters, GV could not determine whether a pair of letters had the same name. Second, she should not access the orthographic structure and meaning of visually presented words, although she could access meaning from orally spelled words and she could access orthographic structure from meaning in written words. Third, GV could access partial semantic information from pictures and Arabic numerals. Based on this pattern of results, we conclude that the form of alexia manifested by our patient results from failure to access the graphemic representations of letters and words from normally processed visual input. The findings further suggest that access to letter forms and grapheme representations are sequentially ordered stages of processing in word recognition. The results also suggest that graphemic processing may be a distinct property of the left hemisphere.

INTRODUCTION

The terms "alexia without agraphia" and "pure alexia" refer to a class of severe reading disorders appearing in the context of (relatively) intact written language production. These disorders are typically associated with lesions of the left posterior cortical and subcortical areas (Benson, 1985: Black & Behrmann, 1994; Damasio & Damasio, 1983; Déjèrine, 1892; Geschwind & Fusillo, 1966). The residual reading abilities manifested by patients with pure alexia may vary extensively. Some patients retain the ability to identify words correctly through a laborious process of recognition of the individual letters. The hallmark of this "letter-by-letter" approach to reading is an increase in reading latencies as a function of

---

Requests for reprints should be addressed to Alfonso Caramazza or Michele Miozzo, Department of Psychology, William James Hall, Harvard University, 33 Kirkland St., Cambridge, MA 02138, USA.

The research reported here was supported in part by an NIH grant NS 34073 to AC. We thank Jeff Anderson and Gedeon Padwa for their assistance in material preparation and test administration and Gail C. Ross for her editorial help. We are profoundly in debt to GV for her inexhaustible patience and good humour over the innumerable hours of testing.

word length. Other patients seem to be completely unable to read words—a form of deficit that is commonly accompanied by a severe impairment in naming letters.

Two approaches have been taken in developing explanations for the different forms of pure alexia. One group of explanations has attempted to characterise the various forms of the disorder in terms of damage to one or more stages of processing in a model of the reading process. Within this approach several proposals have been made. Some authors have argued that the deficit involves relatively early stages of processing (Farah & Wallace, 1991; Friedman & Alexander, 1984; Montant, Nazir, & Poncet, this issue; Rapp & Caramazza, 1991; Sekuler & Behrmann, 1996; and see Behrmann, Plaut, & Nelson, this issue, for a comprehensive review of this proposal). On this view, the deficit is not specific to reading but to all types of visual stimuli (numerals, symbols, pictured objects). The reduced processing efficiency resulting from the perceptual deficit forces the patient to adopt a "divide and conquer" strategy resulting in the letter-by-letter behaviour. Other authors have located the source of deficit at a stage of processing specific to letters (Arguin & Bub, 1993; Bub & Arguin, 1995; Howard, 1991; Kay & Hanley, 1991; Reuter-Lorenz & Brunn, 1990). On this view, the deficit specifically involves the efficient recognition of letters, forcing the patients to adopt a serial letter recognition strategy. Others still have suggested that the deficit specifically involves processing multiple objects simultaneously (Kinsbourne & Warrington, 1962; Levine & Calvanio, 1978). For example, Kinsbourne and Warrington (1962) found that three letter-by-letter readers had normal recognition thresholds for single letters and pictures but abnormal thresholds for multiple stimuli presented simultaneously. And others have argued that all early stages of visual processing of letters are intact and that the deficit arises because of a faulty access to intact lexical orthographic forms (Patterson & Kay, 1982), or because the lexical orthographic representations themselves are damaged (Warrington & Shallice, 1980). Of course, as has been noted previously (e.g. Farah & Wallace, 1991; Friedman & Alexander, 1984; Price & Humphreys, 1992; Rapp & Caramazza, 1991), these explanations of pure alexia are not necessarily mutually exclusive. They could represent correct accounts for different sub-types of the disorder. That is, there may be patients who are alexic because they fail to process normally all types of visual input, others who are alexic because of a selective deficit in recognising letters, and so on.

A different approach to explaining pure alexia has been taken by Coslett and Saffran (1989a, 1989b, 1992, 1994; see also Coslett & Monsul, 1994; Saffran & Coslett, this issue) based on considerations of the neuroanatomical correlates of this disorder. Déjèrine (1892) and Geshwind and Fusillo (1966) had earlier proposed that pure alexia reflects the disconnection of intact visual processes in the right hemisphere from the language areas of the left hemisphere. Coslett and Saffran have added to this claim the assumption that the performance of pure alexics principally reflects the reading capacities of the right hemisphere.

And since a number of pure alexics appear to have considerable reading skills (e.g. Coslett & Saffran, 1994; Shallice & Saffran, 1986), the reading capabilities of the right hemisphere must be equally considerable. In fact, these authors have proposed that the right hemisphere has processes that can support the recognition of the orthographic and semantic content of written words. However, compared to the left hemisphere, the right hemisphere's capabilities are limited in at least two respects: It cannot process abstract words and grammatical morphemes and it cannot process the phonological structure of words (see also Coltheart, 1980; Saffran, Bogyo, Schwartz, & Marin, 1980; Zaidel & Peters, 1981; Zaidel & Schweiger, 1984; but see Patterson & Besner, 1984 for criticism).

Coslett and Saffran (1994) also make two highly specific assumptions concerning interhemispheric connections. One assumption is that only structures that are functionally homologous in the two hemispheres are connected (directly). Thus, for example, the semantic system in one hemisphere is connected only to the semantic system in the other hemisphere and not to any other component. The other assumption is that not all processing mechanisms in one hemisphere are connected directly to their counterparts in the other hemisphere. The privilege of direct connection is restricted to those systems that the authors claim have a "venerable phylogenetic history"—namely, early perceptual processes, and the representations of the structural and the semantic features of objects.

The right-hemisphere hypothesis of pure alexia explains the observed variations in performance among these patients by proposing that the variation reflects differential degrees of integrity of interhemispheric connections. On this view, letter-by-letter readers are those who have intact interhemispheric connections between neural structures dedicated to the processing of objects' visual forms. The left hemisphere can thus interpret the visual form of each letter of a word that it receives from the right hemisphere as a distinct visual object and it can then retrieve the letter names. Once available, the names of the letters are used to recognise the word (in some unspecified way). By contrast, those pure alexic patients who fail to read words altogether are assumed to have damage to the interhemispheric connections that transfer visual information. This form of damage prevents the left-hemisphere reading mechanisms from receiving any input, with the consequence that these patients are unable to read at all.

Coslett and Saffran (1994) also argued that their proposal can explain why some alexic patients perform so well in lexical decision and semantic judgement tasks with written words despite their poor performance in pronouncing them. Their explanation of these contrasting effects goes as follows: Word pronunciation is impaired because this task requires access to left-hemisphere mechanisms which, by hypothesis, are damaged in these patients; however, lexical decision and semantic judgement tasks are relatively spared because they can be performed directly by the right hemisphere.

In this paper, we address some of the issues raised by the different theories of pure alexia by considering the performance of a patient (GV) who presents with a particularly clear set of dissociations as the result of damage to the left posterior cortical areas and the corpus callosum. GV was virtually unable to read and was impaired in reading letters and Arabic numerals as well as in naming pictures. In contrast, her ability to recognise orally spelled words was intact, suggesting that graphemic information is preserved but not accessible through the visual modality. We also replicated and extended the recent finding of Cohen and Dehaene (1995, 1996) of a dissociation between impaired reading of Arabic numerals and preserved recognition of their quantity value. Finally, we were able to show that GV could access partial semantic information from pictures but not from written words. Based on these results, we propose that GV's reading impairment reflects a failure to access the abstract (graphemic) representations that specify the identity of letters and, consequently, failure to access the orthographic form of words and their meaning. A different functional impairment is responsible for GV's deficit in processing pictures and Arabic numerals. With this type of stimuli, access to semantic information is still possible, though limited and insufficient to support the selection of the correct names[1]. We argue that our data are problematic for the hypothesis that there are duplicate word recognition mechanisms in the left and right hemispheres. However, with respect to object recognition, the data are consistent with the hypothesis that some form of rudimentary conceptual knowledge, accessible through visual representations of objects, is encoded in the right hemisphere.

CASE REPORTS

GV is an 84-year-old, right-handed woman with a college education. She worked as a nurse for many years prior to retirement. In October 1994, she had an infarct involving the left posterior cerebral artery. An enhanced CT scan (June 1995) shows a lesion in the left occipital and posterior temporal areas, extending to the corpus callosum (see Fig. 1). An additional area of hypodensity, probably due to a silent, previous CVA appears in the right parietal region. The patient presents with a right visual field cut. The only deficits revealed by the neuropsychological testing were naming difficulties for visually presented stimuli, including objects, pictures, colours, and reading/naming difficulties for letters, numerals, and words. A series of tests was administered to determine whether, as a consequence of her right parietal lesion, there were signs of visual neglect. No evidence of visual neglect emerged

---

[1] Obviously the term "pure alexia" captures only one aspect of GV's impairment—that related to reading. Other terms might be used to describe the additional impairments shown by our patient (e.g. "optic aphasia" for her deficit in processing objects). Nevertheless, the term "pure alexia" captures one aspect of GV's reading impairment—the fact that her reading difficulty is not accompanied by dysgraphia or aphasia.

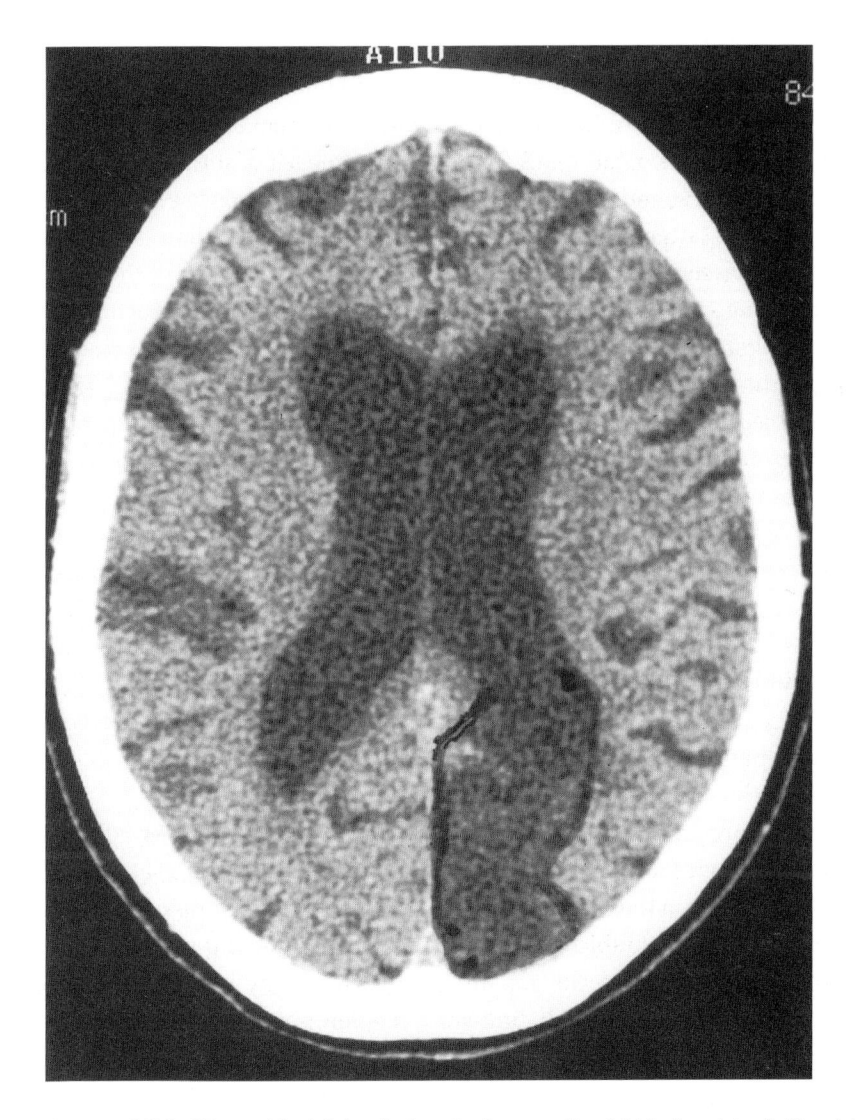

Fig. 1. Representative cut of GV's CT scan (the left hemisphere is shown on the right in the picture). The picture shows a lesion of left occipital and posterior temporal cortical areas extending to the splenium of the corpus callosum. An area of hypodensity is also present in the right parietal region.

from any of the tests (line bisection, various cancellation tasks, drawing from copy and from memory).

The study reported here was started 3 months after the onset of the illness and continued for 6 months. During this period GV's condition remained neurologically and functionally stable. The investigation focused on her abilities to process letters, words, numerals, and pictured objects. The results of the various experimental tasks are presented in separate sections organised as follows: (1) general naming, reading, and spelling performance; (2) visual object processing; (3) letter and word processing; (4) number processing. Each section concludes with a summary of the relevant findings. We begin with a description of the general procedures used in the study.

GENERAL PROCEDURES

GV was asked to perform two types of tasks: (1) tasks in which the stimuli were presented for unlimited exposure, and (2) computerised tasks in which stimuli were shown for limited duration. The procedures in these two types of tasks were slightly different, and hence they will be described separately.

Time-unconstrained Tasks

In these tasks, two types of responses were produced—either a naming or a nonverbal response (e.g. pointing). In the naming tasks, the patient's last response to each item was scored; this procedure acknowledges the possibility that the patient's initial responses were attempts to approach the correct response. With tasks requiring nonverbal responses, the first response was retained for further analyses. (It should be stressed, however, that the pattern of results remained essentially unchanged with different scoring procedures.) Whenever the same list of stimuli was administered multiple times or was used in different tasks, order of item presentation was counterbalanced. Each administration of the experiment proper was preceded by a practice block. Alphanumeric stimuli were printed in 34-point Helvetica font. (Essentially the same pattern of performance was found when stimuli were presented in a smaller size.) The size of lower-case letters was normalised to the size of upper-case stimuli. Only letters judged to have different shape in upper and lower case (Boles & Clifford, 1989) were used in tasks where letters were shown in different formats. Specifically, we selected the following pairs of letters: *A-a*, *B-b*, *D-d*, *E-e*, *F-f*, *G-g*, *H-h*, *N-n*, *P-p*, *Q-q*, *R-r* and *T-t*.

Time-constrained Tasks

The patient was seated at a distance of approximately 45cm from a computer screen. A trial had the following structure: The patient initiated a trial by pressing the space bar; then a fixation point (a cross) appeared for 700msec, which was immediately replaced by the stimulus. Because of GV's visual cut, targets were always shown at the left of fixation. To respond, the patient pressed one of two keys with either the right or left hand, on the right and left of the body midline, respectively. Pilot tests revealed that with this procedure the patient did not miss any stimulus. The assignment of keys to type of response was counterbalanced across testing sessions. Stimulus presentation and recording of response latencies were both controlled by PsychLab software (Bub & Gum, University of Victoria, Victoria, British Columbia, Canada). In order to obtain a substantial number of observations and to reduce response variability, the computerised tasks were administered multiple times across several testing sessions. The order of presentation of the experimental blocks was varied systematically for each administration of the test. Correct response latencies exceeding three standard deviations were replaced by their condition's mean. To have an equal number of responses in each cell, the latter procedure also was adopted with erroneous responses. Different sorts of stimuli were shown via computer presentation: pictures, letters, words, and digits. The maximum visual angle of pictures was 5°. Alphanumeric characters were in 34-point Helvetica font.

NAMING, READING AND SPELLING

The results presented in this section aim to establish the selectivity of GV's naming deficit. Specifically, we investigated whether her impairment is restricted to the visual modality and whether, within this modality, it extends to different sorts of stimuli (pictures, colours, letters, words, and numerals). Comparisons of her naming abilities across modalities were used to determine the intactness of semantic knowledge and of the mechanisms for lexical production. A deficit in naming letters and numerals might originate from a difficulty in accessing the phonological form of the stimuli. To examine this hypothesis, we administered tasks that required the oral production of letter names (oral spelling) or number names (e.g. retrieval of numerical facts). The analysis of her performance in various spelling tasks also provides the basis for drawing conclusions about the integrity of her knowledge of a word's orthographic representation.

Picture Naming

GV's ability to name pictures was formally tested by means of Snodgrass & Vanderwart's (1980) picture set ($N = 260$). This set was administered twice. Response modality varied in the two administrations (oral vs. written). As shown in Table 1, GV was severely impaired. She was able to name only about 50% of the pictures successfully. Furthermore, her level of accuracy remained unchanged across output modalities. The majority of her erroneous responses can be broadly classified as *semantic confusions*, which account for 82% and 71% of her errors in oral and written naming, respectively. Within this category of responses we included semantic substitutions (e.g. *table* → "seat", *tie* → "belt"), superordinate names (e.g. *necklace* → "piece of jewellery", *pepper* → "vegetable") and functional descriptions of the objects (e.g. *saw* → "used for chopping wood", *ruler*, → "to take measures"). Visually similar errors (e.g. *balloon* → "flower", *clock* → "button") and perseverations were produced far less frequently.

Table 1. Oral and Written Picture Naming—GV's Responses

		No. of Responses/Task			
		Oral Naming		Writing Naming	
Type of Response	Example	No.	%	No.	%
Correct		125	(48)	132	(51)
Errors					
Semantically related	ear → "nose"	111	(42)	91	(35)
	pepper → "vegetable"				
Visually similar	balloon → "flower"	10	(4)	12	(5)
Perseverations		13	(5)	19	(7)
Others	toaster → "stairs"	1	(1)	6	(2)

Colour Naming

GV's severe naming deficit extended to colours. She successfully named the colour of only 4/12 crayons. All errors consisted of colour name substitutions.

Naming in Nonvisual Modalities

In naming a set of objects shown for tactile exploration (from Hillis & Caramazza, 1995a), GV responded correctly to 46 out of 47 items (98%). She also performed well in naming orally presented definitions of the same objects (96% correct). However, in naming pictures of these objects she was correct only 62% of the time. Her errors (17/18) were almost all semantic confusions (e.g. *sock* → "belt", *spatula* → "a kitchen thing"). The difference among tasks is statistically reliable [Cochran, Q(2) = 28.7, P < .001]. The contrast between the performance in the visual confrontation and the other naming tasks clearly demonstrates the modality/specificity of her naming disturbance. Furthermore, GV's good performance in the tactile and naming-to-definition tasks excludes the possibility that her visual naming deficit results from damage to the semantic system.

Letter Naming

In several testing sessions, GV named each upper- and lower-case letter of the English alphabet 20 times. Upper- and lower-case letters were presented in separate blocks of 26 stimuli each. GV was impaired in naming the stimuli in both formats, although her accuracy was slightly higher with upper- compared to lower-case letters [44% vs. 30%; $\chi^2(1) = 10.7$, $P < .001$]. This asymmetry, also noticed in other patients with impaired naming of letters (e.g. Perri, Bartolomeo, & Silveri, 1996), presumably reflects differences in the degree of confusability among letters in the two types of formats. Correct responses were produced for a restricted group of letters—the characters *A, C, E, I, O, U,* and *i, m, o, u* accounted for about 40% of correct responses in each set. The probability with which individual letters were produced as incorrect responses was also unequal, reflecting a tendency to perseverate with a subset of letter names.

Reading Aloud

GV was virtually unable to read words. Presented with a set of 70 words controlled for frequency and length (Goodman & Caramazza, 1986), she correctly read only one word, *chair*. Six months later, her performance remained unchanged (0/70 correct). GV's attempts to read were very laborious. Usually, she started by trying to identify single letters, then tried to concatenate them into syllables, sometimes producing a word (e.g. *ruin* → "guru"). Occasionally, responses contained parts of the stimulus (e.g. *chipmunk* → "check" or *special* → "deal"). Perseverative responses were also noticed.

Naming of Numerals

Different lists of numerals were used. In one task, each digit in the range 0–9 was presented, in a random order, 10 times. GV successfully named 60% of the items. Correct responses were unequally distributed across target digits; for example, *8* was named correctly nine times while *3* was named correctly only three times. She had a bias to respond with the digits *4*, *5* and *6*, which accounted for 72% of her error responses. Her level of performance decreased dramatically with 2- to 5-digit numerals: She correctly named only 29/280 (10%) items overall, and virtually none of the 4- and 5-digit numerals (see Table 2). In this task, numerals were presented without commas, and the patient was instructed to read each stimulus as a "single number", rather than to read the digits

Table 2. Naming of 1- to 5-digit Arabic Numerals—GV's Correct Naming and "Syntactic" Errors

Target	Correct No.	%	Syntactic Errors No.
1-digit	17/40	(42)	5
2-digit (teens)	12/40	(30)	4
2-digit	10/40	(25)	1
3-digit	6/40	(15)	0
4-digit	1/40	(5)	0
5-digit	0/120	(0)	6

sequentially. Errors typically consisted of digit substitutions (e.g. 647 → "827", 39087 → "49021"). "Syntactic errors"—characterised by the deletion (e.g. 647 → "67" or "6") and/or addition (e.g. 647 → "1647" or "64007") of one or more digits—were observed very rarely (< 1%).

An analysis was carried out to determine whether GV's naming errors tended to be "close" in magnitude to the expected response (e.g. *3* → "4" or *8* → "7"). We analysed a large corpus of errors ($N = 369$; details about these errors will be presented later in the Magnitude Judgement Task). There were only eight expected responses in the corpus used for this analysis (the digits *0* and *5* were never presented). The distribution of the observed errors as a function of their numerical distance from the target was contrasted with the distribution expected by chance (i.e. when each digit has an equal probability of being selected; Cohen & Dehaene, 1996; see Fig. 2). Responses that were one or two digits larger/smaller than the target were considered as "close errors". The observed frequency of close errors was not significantly different from that expected by

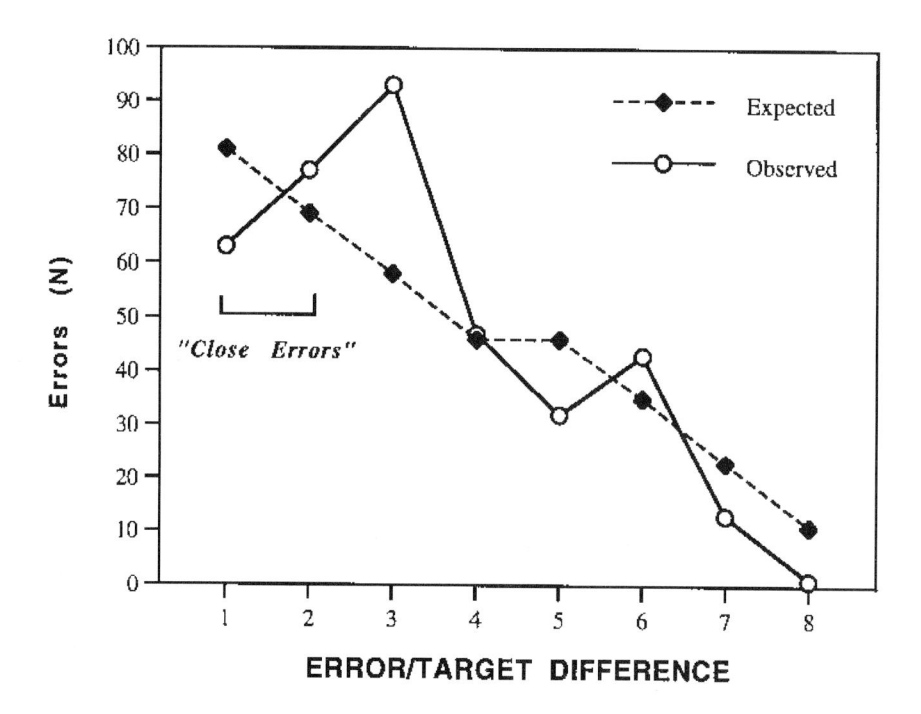

Fig. 2. Distribution of GV's errors (N) as a function of their numerical difference from the target (solid line) in the Oral Naming Task with Arabic digits. The figure also shows the distribution expected if responses were randomly selected (broken line). Note that the digits 0 and 5 were never presented in this task.

chance [$\chi^2(1) < 1$]. This finding indicates that numerical distance is not among the factors significantly affecting the selection of the erroneous responses in this task.

Numerical Facts

GV was asked to retrieve various numerical facts such as "how many pennies are in a dollar?" or "what is the year of the discovery of America?" She performed flawlessly with oral questions of this sort. Excellent performance was also found with questions concerning personal numerical information (e.g. her year of birth). Further evidence that access to number names was intact is provided by her ability to solve simple calculations (e.g. 2 + 3; 6 × 7) with orally presented stimuli, These results indicate that her number naming deficit is not due to damage to lexical production mechanisms or the number semantic system.

Aural Recognition of Orally Spelled Words and Spelling Tasks

The list of words used to assess reading was also used in a recognition task of orally spelled words and in written and oral spelling-to-dictation tasks. As shown in Table 3, GV performed flawlessly in recognising orally spelled words and made very few errors (4.3%) in the two tasks. The few errors (6/140) all involved

Table 3. Spelling Tasks—GV's Correct Responses

Task	Correct[a] No.	%
Written spelling	66	(94)
Oral spelling	68	(97)
Aural recognition of orally spelled words	70	(100)

[a]Total = 70.

low-frequency words and were phonologically plausible realisations of the target (e.g. *pirate* → PYRATE, *pigeon* → PIGION), with one exception (the omission of "t" in *instinct* → INSTINK). The results of these tasks indicate that GV has normal access to the orthographic structure of words both in recognition (aural recognition of orally spelled words) and in production (spelling).

Discussion

The results of the tasks reported in this section reveal the extent and severity of GV's naming impairment with visual stimuli: Naming was severely disrupted for pictures, colours, letters, words, and numerals (see summaries in Tables 1 and 2). Her contrasting performance in tasks with nonvisual stimuli allows us to exclude some hypotheses about the functional level of her deficit. Specifically, because GV successfully named objects from tactile exploration and from verbal description, we can exclude as causes for her visual naming deficit both damage to the semantic system and damage to the word production mechanisms. We can also exclude a deficit in producing the names of letters, since she could retrieve them normally in the Oral Spelling Task. Similarly, since GV successfully named numerical facts in response to orally presented stimuli, we can infer that she has intact access to the lexical forms of numbers. GV's normal ability to recognise orally spelled words shows that her orthographic input lexicon is intact (Caramazza & Hillis, 1990; though see Warrington & Shallice, 1980, for a contrasting viewpoint).

VISUAL OBJECT PROCESSING

The tasks reported in this section address three points. First, a comparison of the results with various visual processing tasks allows the identification of the locus of functional damage responsible for GV's impairment with visual stimuli. We address whether such damage arises (a) at the level of perceptual analysis, (b) at the level of access to information about the shape of familiar objects (structural description), or (c) at the level of semantic access. Second, we attempt to determine whether semantic information is fully accessed with pictures. (In a later section, the accessibility of semantic information with pictures will be contrasted to that found with visually presented words.) Third, since several of the tasks presented in this section have previously been given to other patients with alexia, the results of these tests allow a direct comparison of the performance of our patient with that of previously documented cases.

Visual Perceptual Tests

The percentage of correct responses produced by GV in a series of visuoperceptual discrimination tasks from the BORB (Riddoch & Humphreys, 1993) is comparable to that of a group of 39 neurologically intact control subjects (age 50–80; see summary in Table 4). These tasks require very fine discriminations of line lengths, line orientation, size of squares, and the position of small gaps in circles. GV's good performance in these tasks rules out a low-level perceptual deficit as the main cause of her modality-specific impairment. In addition, her ability to trace the outlines of overlapping stimuli indicates that she can segregate visual objects.

Matching Objects from Different Views

The structural description of an object is a mental representation specifying the visual features of a canonically oriented familiar object (e.g. Riddoch & Humphreys, 1987a). Access to these representations can be evaluated by means of a recognition test in which one has to decide whether the same object is pictured from different perspectives. The material and the norms for this task were from Riddoch and Humphreys' (1993) BORB test. Subjects were presented with three pictures: two of the pictures represented different views of the same object; the third picture, a foil, showed an object visually similar to one view of the target. GV's accuracy (46/50; 92%) did not differ from that of controls (age 50–80; 89.8%).

Reality Decision

GV was asked to discriminate between real and chimaeric visual objects. The latter objects were created by replacing one part of one object with a part from another object of the same category (e.g. the head of a donkey was mounted on the body of a dog). Stimuli ($N = 64$) and norms were from the BORB (Riddoch & Humphreys, 1993). GV performed within the normal range: 83% vs. 81% correct for GV and the mean of controls, respectively. The comments occasionally made by the patient revealed that she was able to identify stimuli even when she failed to name them. For example, she correctly accepted the picture of a *pig* as real, but was only able to describe it as "a farm animal," and although she correctly rejected the chimaeric snake/turtle as unreal, she described it as having "the head of a goat." Her good performance in the Reality Judgement Task, paired with her good ability to match objects from different views, provides convincing evidence that GV can successfully access the stored structural descriptions of objects.

Picture Categorisation

GV performed flawlessly in sorting 30 pictures into the categories animals vs. plants. She also correctly distinguished (48/52; 92%) between

Table 4. Visual Discrimination Tasks—GV and Controls

Feature	% Correct GV	Controls	z-scores
Line length	80	89	−1.8
Line orientation	90	82	+0.8
Square size	83	91	−0.9
Gap position	82	87	−0.5

pictures of food (e.g. *sandwich*, *apple*) and non-food items, including kitchen utensils.

Picture Association (Visually Similar Foils)

In this task, GV selected the two associated pictures out of a set of three. The distractor was visually similar to one of the targets (e.g. *pear* was presented with *light bulb* and *light switch*). The stimuli were from Hillis and Caramazza (1995a). GV correctly matched 32/33 (97%) of the items. In a separate session, she named only 34% of the pictures correctly.

Picture Association (Related Foils)

Picture categorisation and picture association tasks have previously been used to evaluate the intactness of semantic access (e.g. Coslett & Saffran, 1989b; Iorio, Falanga, Fragrassi, & Grossi, 1992; Manning & Campbell, 1992). Consistent with the interpretation reached by these earlier investigators, we should conclude that GV's access to the conceptual representation of objects is intact. However, such a conclusion would be unjustified if based only on such performance, since access to incomplete conceptual representation of objects might be sufficient to support good performance in the kind of recognition tasks presented earlier (e.g. Hillis & Caramazza, 1995a). Access to an intact (complete) semantic representation is not needed for the rather coarse discrimination between animals and plants; and it is not needed for the finer-grained edible vs. inedible discrimination either. The retrieval of an incomplete representation that includes the right features is sufficient for accepting an object as edible (e.g. <tasteful>, <found in a grocery store>, <for sandwiches>) or inedible (e.g. <it has wheels>, <electrical appliance>, <made of metal>). Analogous arguments can be made to account for good performance in the triadic picture association task. Correct matching in this task may simply result from a differential judgement: The associated items share many more features with each other than either one does with the unrelated item. A judgement of this sort does not require access to a complete semantic description. All that is needed is that associated objects should activate a large enough number of common features to distinguish between related and unrelated items. In the tasks used in previous researches, the foil were always semantically unrelated to the other two items. Therefore, the foils very rarely activated semantic features in common with the two other items, a fact that should guarantee a correct match even in conditions of partial disruption of semantic access. Thus, the results presented do not exclude the possibility that GV has a deficit in accessing semantic representations. And, in fact, when GV is given tasks that require access to detailed semantic information, her performance deteriorates dramatically.

Hillis and Caramazza (1995a) introduced the following modification to the triadic matching task: The visually similar foil was replaced by a related, though less associated, foil. To illustrate, the related pair *light bulb* and *light switch*, which in the original task was shown with the visual foil *pear*, was now shown with the semantically related foil *traffic light*. GV's performance in the task with semantically related foils differed dramatically from the original triadic association task: In the new task, she responded correctly on only 15/33 trials (45%; compare this to 97% correct in the previous task). The lower performance in the related foils task is not because there is ambiguity as to the correct choice for the new items. When GV was given an aural version of this task (i.e. she was given the names of the objects in a triad), she performed flawlessly (33/33). Thus, we have evidence that GV's semantic system is intact but that access to representations through the visual modality (i.e. for pictures) is impaired.

Property Judgement Task

GV answered yes/no questions ($N = 160$) about specific semantic features of objects (e.g. *Is it (lemon) sour? Does it (eagle) sing?*). For each item there were two "yes" and two "no" spoken questions. A wide range of features were queried: visual, functional, associative, and encyclopaedic. Care was taken to exclude questions whose answers could be derived from features contained in the drawings. In one condition a picture was shown, in the other the name of the picture was orally presented. A striking dissociation was found between conditions: When targets' names were presented orally she was almost invariably correct (97%); with pictures, however, she responded correctly only 83% of the time [*McNemar, P < .001*].

Spoken Word/Picture Verification

In this task (from Hillis & Caramazza, 1995a), GV was asked to recognise whether a spoken name matched a given picture. Pictures from the Snodgrass & Vanderwart (1980) set were presented with three different words on three separate occasions: the name of the picture, a semantically related noun (e.g. grapes–*cherry*, guitar–*violin*), and a phonologically similar noun (e.g. eagle–*needle*, foot–*flute*). An item was scored as correct only if the subject accepted the correct name and rejected the two foils. GV performed rather poorly in this task: She responded correctly only 64% (167/260) of the time. The vast majority of her errors (81/96; 84%) involved the acceptance of a semantic foil. The few times (8/96; 8%) in which GV accepted a phonological foil, the foil was visually similar to the target (e.g. anchor → "hanger"). Thus, we have converging evidence that GV's performance is impaired when the task requires her to make fine semantic distinctions.

Discussion

The results reported in this section seem to exclude a visual perceptual deficit as the basis for GV's poor naming performance. No indications of damage to the visual processing system emerged from tasks requiring very precise discriminations of perceptual features such as orientation, length, size, and position (see Table 4). Furthermore, she performed flawlessly in object recognition tasks (e.g. the Reality Decision Task), which require excellent perceptual abilities for normal execution. Nevertheless, one could object that these tasks were not particularly stringent. More stringent tests of GV's ability to process visual stimuli are presented later in the paper.

Two other important findings are reported in this section (see Table 5 for summary): (a) intact performance in tasks requiring access to structural knowledge of visual objects, and (b) impaired performance in visually accessing semantic information. The latter findings invite the conclusion that up to the point of access of stored structural descriptions of objects, visual processing is intact; it is the subsequent access to semantic information that is disrupted. However, access to semantics is not completely impossible, but partially limited. This characterisation of her ability to access semantics through visual modality is in agreement with the type of errors she commonly produced in picture naming—semantic substitutions. Errors such as *ruler* → "to take measures" or *fox* → "wild animal" reveal that despite her failure to retrieve the entire set of target's semantic features, some of these features were accessed successfully. Note, however, that GV's deficit cannot be attributed to damage to semantic knowledge itself, since she performed naming and other semantic tasks normally when the input involved the auditory and the tactile modalities.

Several of the tasks presented in this section are identical to those given to other patients with an "isolated" right brain for visual processing (patients with selective damage of the left posterior areas and of the corpus callosum), including some of the patients with alexia reported by Coslett & Saffran (1989b, 1992). These cases all tend to show intact access to the structural representations of objects (see Iorio et al., 1992 for review). In studies where semantic access from visual objects was examined extensively (e.g. De Renzi & Saetti, 1997; Hillis & Caramazza, 1995a; Iorio et al., 1992; Riddoch & Humphreys, 1987b), the results are identical to those observed with GV: Access was only partial.

In several other studies the semantic tasks used (e.g. picture classification) did not involve the discrimination of detailed semantic information of visual objects. Thus, it is not possible to make direct comparisons across all relevant cases on this aspect of language processing. However, some of the tasks used in the latter studies (e.g. picture association with visual foils) were also used with GV (and patient DHY studied by Hillis & Caramazza, 1995a). For these tasks, similar results were found across studies: Patients performed well. However, as we have seen in GV, these tasks are not sufficiently sensitive to reveal an impairment of semantic access with visual objects. Therefore, it cannot be excluded that in previously documented cases (including those reported

Table 5. GV's Performance in Semantic Tasks—Summary

Task	% Correct Pictures	% Correct Words
Categorisation task		
Animals vs. Plants	100	100
Food vs. Nonfood items	92	96
Association task		
Visual foils	97	100
Related foils	45	100
Property judgement	83	87
Word/Picture verifications	64	–

by Coslett & Saffran, 1989a, 1992) complete semantic access from visual objects was impaired.

WRITTEN LETTERS AND WORDS

In the preceding section, we investigated the integrity of access to the structural representation of familiar objects. We now investigate whether analogous structural information is available for written letters. We also examine the availability of information about abstract letter identity (graphemes) and the extent to which GV can access information about the orthographic and semantic content of written words. In designing the latter tasks we followed the procedures described by Coslett & Saffran (1994).

Reality Decision

GV was asked to distinguish between familiar upper/lower-case letters ($N = 52$) and pseudoletters ($N = 38$). The latter are stimuli that look like letters; they were created by rearranging the strokes of letters of the Roman alphabet (see examples in Fig. 3). GV correctly classified all the stimuli, despite the fact that she frequently produced the wrong name when she spontaneously named the letters (it was not required by the task; incorrect naming was also observed when GV performed the Reality Judgement Task with objects).

Orientation Decision

This task, devised by Cooper & Shepard (1973), was used to examine the availability of infor-

Fig. 3. Examples of pseudoletters used in the Reality Decision Task.

mation about the canonical orientation of letters and digits. The task requires the discrimination between normally orientated and reflected characters. Stimuli were presented in each of six orientations, equally spaced by 60° starting from 0° (see examples in Fig. 4). The asymmetrical letters and digits *R, J, G, 2, 5* and *7* were used in this task. Each stimulus appeared twice in each orientation, once in the canonical and once in the noncanonical (reflected) orientation. As is evident from Fig. 4, GV's performance in this task is indistinguishable from that of an age-matched control subject (85% and 87% correct judgements, respectively). Such a good level of performance cannot have resulted from the fortuitous selection of letters and digits easily identifiable by the patient. In a separate session, she successfully named only 15% of the stimuli[2]. The results of the orientation task provide support for the hypothesis that GV has access to the structural representations of letters *and* digits.

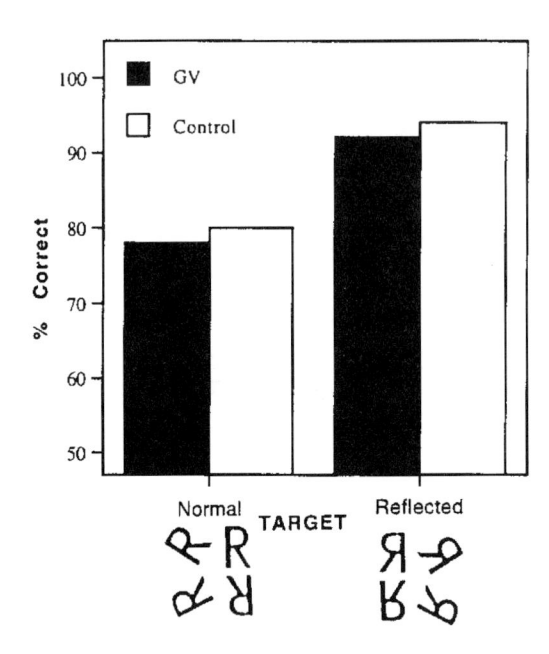

Fig. 4. Correct performance (%) for GV and a matched control subject in a letter/digit orientation task (Cooper and Shepard, 1973). Examples of normally oriented and "reflected" stimuli are show.

Same/Different Name Decision Task

The letter matching task devised by Posner and Mitchell (1967) was given to GV. In this task, the response "same" is expected with pairs of letters that are either physically identical (PI; A–A, a–a) or name identical (NI; A–a, B–b). When the two letters had a different name (D), the subject was instructed to respond "different." On 50% (N = 288) of the trials, D pairs were shown; the remaining trials consisted on an equal number of PI and NI pairs. Letters were shown one at the time for 200msec, separated by a 400msec inter-stimulus interval. On each trial, the stimuli appeared at two distinct locations—in the left quadrant above or below fixation. The order of presentation was counterbalanced between the two locations. As illustrated in Fig. 5, the patient responded accurately only with PI pairs (93% correct); with NI and D pairs her performance was much less accurate (63% and 61% correct, respectively) although above chance [$\chi^2(1)$ = 5.1 and 6.7; for both, $P < .001$]. These results indicate that access to information about grapheme identity is severely impaired.

Letter Transcoding

Additional evidence on the availability of information about grapheme identity can be obtained with a task in which visual letters are to be reproduced (written) in a different format (e.g. $A \rightarrow a$, $d \rightarrow D$). GV successfully transcoded 31/84 (37%) of the upper/lower-case letters (only the 14 characters with different shapes in the two formats were used). Errors consisted of well-formed but incorrect letters (e.g. $A \rightarrow p$).

The results of the letter (and digit) tasks presented thus far support the conclusion that GV has normal access to the structural descriptions of letters (and digits) but not to information about abstract letter identities or graphemes.

---

[2] (Opposite). The reliability of these results is further supported by the findings of another Orientation Judgement Task. In this task, almost the entire set of upper- and lower-case letters was used (N = 41). Each character was presented in two orientations (canonical and noncanonical). Noncanonical letters were "upside-down" (rotated 180° on the fronto-parallel plane) or "reflected." GV always discriminated the stimuli correctly.

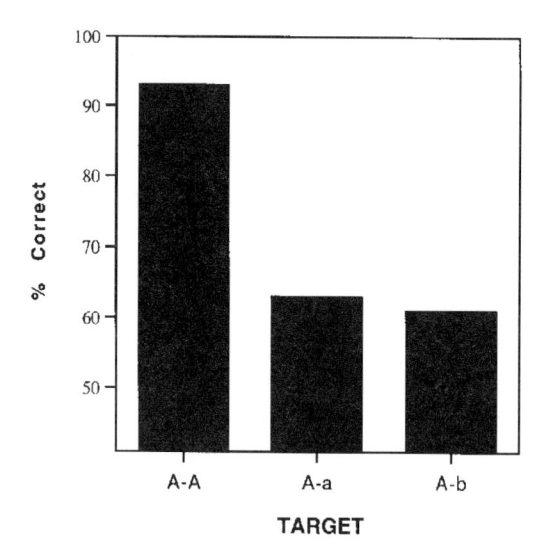

Fig. 5. GV's correct performance (%) in the Same vs. Different Name Decision Task (after Posner & Mitchell, 1967).

Lexical Decision Task

Successful discrimination of familiar vs. unfamiliar words is based on the retrieval of appropriate lexical orthographic information. Lexical decision performance can thus reveal the extent to which orthographic information is available. GV was shown a set of 20 high-frequency words and 20 nonwords. The latter were created by changing one letter in very familiar English words (e.g. *home → HAME*). All stimuli were four letters long and were printed in upper case. GV was asked to decide whether the string of letters formed a word. She was also given strict instructions not to name the stimuli. She was unable to distinguish between words and nonwords (55% correct). This result invites the conclusion that GV cannot access the orthographic lexicon from visual input.

Word Categorisation Task

A word categorisation task (animal vs. artefact) was used to assess GV's ability to access meaning from print. Stimuli were three- and four-letter, highly familiar words. The stimuli were presented for unlimited time on separate cards. GV was asked to sort them into two piles. She was required not to name the words, but to "concentrate" on their meaning. Her level of accuracy in this task does not differ from chance (23/40, 57% correct responses), a result clearly indicating severe difficulty in accessing the meaning of words.

Priming Task

Coslett & Saffran (1989a, 1989b) demonstrated that the reading performance of patients with pure alexia may improve when stimuli are presented for limited exposure. We explored whether GV's performance might similarly improve if written stimuli were presented very briefly. We used a picture categorisation task (animals vs. artefact) that used briefly presented words as primes. If GV can access the meaning of words, her categorisation latencies would be expected to be faster with identical and related (same-category) primes than with unrelated primes. In a separate experiment, pictures were used as primes. The latter experiment served as a control to determine whether a priming effect could be obtained with primes other than words.

The results of the picture categorisation tasks discussed in the previous section convincingly demonstrate that GV can access sufficient semantic information to categorise objects. We, thus, expect that GV would show picture/picture semantic priming. However, given previous results with words, which have shown failure to access lexical representations, we do not expect GV to show semantic priming in the word/picture condition. Picture and word primes were used in separate experiments. While targets always appeared for 200msec, prime presentation varied: 200msec for pictures, 2sec for words. We chose this particular presentation rate for words because at a shorter exposure the patient reported being unable to "see" them. The inter-stimulus interval was set at 600msec for both conditions. Primes and targets were shown in the same position on the screen, at the left of fixation. Word primes were three and four letter long. Target pictures ($N = 36$) appeared once per condition (identical, related, and unrelated), and were equally distributed in the two categories (animals and artefacts). Each priming task was completed three times, separated by several weeks.

GV's response latencies in the different tasks and conditions are presented in Fig. 6. Her performance was highly accurate (90% correct), and did not very as a function of the type of prime (pictures = 90%; words = 89%) or the type of prime–target relation (identical = 91%; related = 89%; unrelated = 88%). Furthermore, her responses were always very fast (mean = 650msec). For picture primes, an ANOVA revealed an effect of prime–target relation [$F(2,35) = 33.2, MS_e = 15,656, P < .001$]. As demonstrated by paired comparisons, her responses were faster, relative to unrelated primes, both for the identical [$t(35) = 8.1, P < .0001$] and the categorically related primes [$t(35) = 5.6, P < .0001$]. In addition, she responded faster to pictures preceded by the identical than related primes [$t(35) = 2.8, P < .01$]. For word primes, an effect of prime–target relation was also observed [$F(2,35) = 4.1, MS_e = 3226, P < .05$]. However, this difference is due to the slightly *longer* responses (32msec) obtained with identical than with related and unrelated primes.

To summarise, a large facilitatory effect was observed with picture primes, but there was no trace of facilitation with word primes. The results of the priming experiments confirm and further extend those of the Word Categorisation Task, in which it was shown that GV does not have access to the semantics of visual words. However, GV does have access to some semantic information from pictures, as indicated by her good performance in the categorisation task and by the large facilitation effect induced by picture primes.

Discussion

Two important findings—one with letters, the other with words—are reported in this section (see Table 6 for summary). With visually presented letters, GV showed a striking dissociation: She performs normally in accessing knowledge of letter shape (e.g. the Cooper & Shepard Rotation Task), but cannot access grapheme identity knowledge. Thus, for example, she could recognise whether a letter was normally oriented, but she was unable to indicate whether two letters had the same name. With visually presented words, there was no indication that GV could access infor-

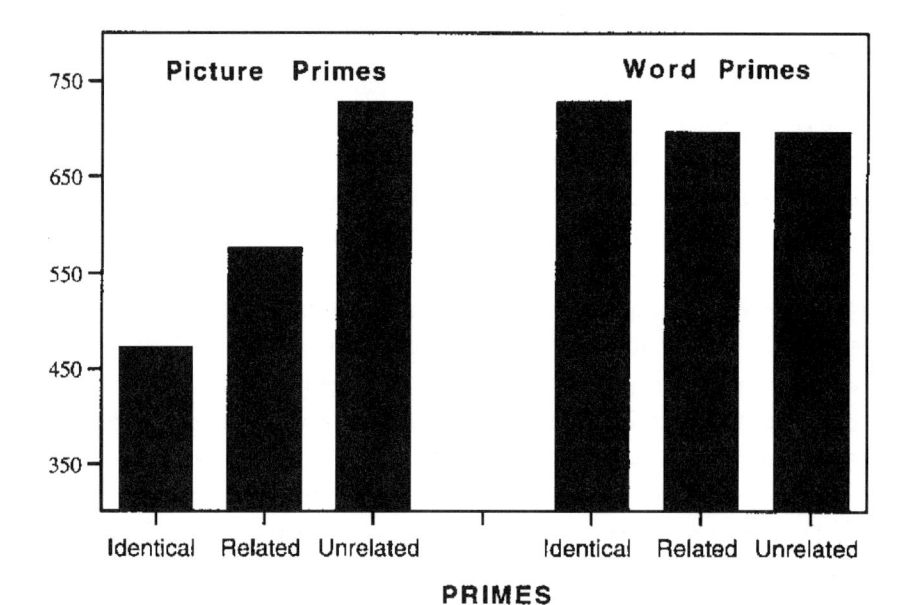

Fig. 6. GV's distribution of response latencies in the Picture Categorisation Task as a function of the type of prime (words vs. pictures) and of the prime–target relation (identical, categorically related, and unrelated primes).

mation about their lexical status or meaning. This conclusion is supported by the results of forced-choice tasks, in which she performed at chance in discriminating words vs. nonwords and animal vs. artefact words. And there was not the slightest indication that GV could access meaning when written words were used as primes in the Picture Categorisation Task. Together, the results with letters and words allow a precise characterisation of GV's reading deficit. It seems to originate from a failure to access graphemic representations from intact letter shape information. And, if graphemic information is obligatory to access the orthographic and semantic content of visually presented words, then it is not surprising that GV failed in reading tasks. Before considering the implications of these results, we present a final set of data concerning number processing that further help determine the nature of GV's impairment in processing visually presented stimuli.

Table 6. GV's Performance in Tasks with Visually Presented Letters and Words Summary

Task	% Correct
Letters	
Reality Decision	100
Transcoding Task	37
Words	
Lexical Decision[a]	55
Categorisation[a]	57

[a]Chance = 50%

NUMBER PROCESSING

The investigation of GV's number processing performance was motivated by a series of findings recently reported by Cohen and Dehaene (1995; see also McNeil & Warrington, 1994). They demonstrated that patients with pure alexia may retain the ability to make magnitude judgements with Arabic numerals despite their difficulties in naming them (for similar dissociations in patients with other patterns of reading impairments see Cipolotti & Butterworth, 1995; Cipolotti, Warrington, & Butterworth, 1995; Cohen & Dehaene, 1996; Cohen, Dehaene, & Verstichel, 1994; Dehaene & Cohen 1996). To illustrate, a patient might fail to name the digit 7 but still be able to recognise that 7 is larger than 5 and smaller than 8. Does GV show a similar dissociation between naming and magnitude judgement? In other words, can we show that GV is able to access magnitude information despite her inability to name digits? Evidence on this issue would further constrain our interpretation of GV's ability to process alphanumerical characters.

Magnitude Judgement (Fixed Standard)

Studies that have examined magnitude judgements with neurologically intact subjects have reliably obtained a clear distance effect. Moyer and Landauer (1967) observed that when subjects decide whether an Arabic digit is smaller or larger than 5, their response latencies are inversely correlated with the distance from the standard 5 (see also Banks, Fujii, & Kayra-Stuart, 1976; Hinrichs, Yurko, & Hu, 1981). Does GV show a similar distance effect in magnitude judgements?

This experiment was an exact replication of the paradigm used by Moyer and Landauer (1967). Each digit in the 1–9 range (with the exclusion of 5) was presented 224 times. Stimuli were shown in blocks of 32 trials, with the following constraints: The same response was not expected for more than 3 consecutive trials and the same digit was not repeated in the successive 3 trials. Targets were shown for a maximum of 2sec and were removed as soon as the subject responded. She was instructed to press separate keys for "larger" and "smaller." GV's performance was surprising accurate: On average, 98.1% of her responses were correct (range across digits: 95%–99%). The distribution of her response latencies as a function of the distance from the standard is reported in Fig. 7. Two aspects of the latter results are worth stressing: First, her responses are extremely fast (mean = 635msec) and, second, latencies tend to decrease with the distance from 5 ($r = .124$, $P < .001$).

To ensure that GV was still impaired in naming digits despite the massive exposure to these stimuli, her naming accuracy was assessed following each administration of the Magnitude Judgement Task. Across sessions, a given digit was named a total of 110 times. GV correctly named 58% of the digits (an analysis of the errors produced by the patient in this task was presented earlier in the paper; see the Naming, Reading, and Spelling section and Fig. 2). Her level of accuracy was almost identical to that in the Digit Naming Task performed before the administration of the Magnitude Judgement Task (54%).

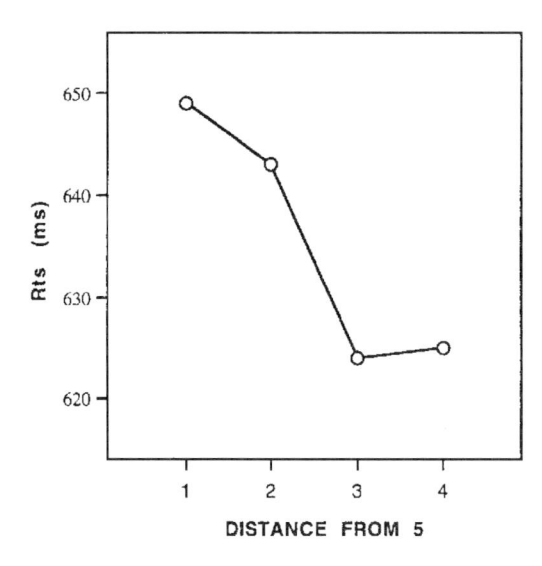

Fig. 7. GV's distribution of correct response latencies as a function of the target's distance from 5 in the Magnitude Judgement Task (Moyer & Landauer, 1967). Targets were Arabic digits.

In a second judgement task, all numerals from 1 to 99 were presented for comparison with the standard of 55 (not presented). Each numeral was shown separately for an unlimited time, and the subject expressed her judgement by saying "larger" or "smaller." She correctly judged 89/99 (90%) numerals, a level of performance that is far more accurate than would be expected from her oral naming performance of a subset of the same stimuli (24/93; 26%).

Magnitude Judgement (Pairs of Numerals)
To explore GV's ability to process visually presented numerals further, she was asked to indicate which of two Arabic digits was larger. In this task, neurologically intact subjects show a distance effect (e.g. Banks et al., 1976; Sekuler, Rubin, & Armstrong, 1971): Response latencies decrease as the numerical difference increases. Thus, for instance, subjects are typically faster at deciding that 3 is bigger than 1 than at deciding that 3 is bigger than 2. An additional effect, the so-called *min* effect, is also repeatedly observed (e.g. Buckley & Gillman, 1974; Parkman, 1971): For constant distances, comparison latencies vary as a function of the value of the smaller digit in the pair. Specifically, responses tend to be faster with pairs having a smaller (e.g. *1–2*) than a larger *min* (e.g. *8–9*). We adopted the experimental procedure used by Banks et al. (1976). The difference (*split*) between the digits of a pair was either 1 or 2. The following digit pairs were presented an equal number of times (24 each): *1–2, 2–3, 1–3, 7–8, 8–9,* and *7–9*. To avoid having the digits *3* and *7* always associated with the responses "larger" and "smaller," respectively, the pair *3–7* was also included (it appeared twice as frequently as any of the other pairs). The two digits of a pair were shown simultaneously, side by side in a horizontal row, 1.5cm apart. Left/right positions were counterbalanced for each digit. GV responded by pressing the key on the same side of the larger digit. In this difficult task, GV responded accurately (correct responses = 98%) and rapidly (mean RT = 928msec). Strong and reliable distance effects were obtained (see Fig. 8): Responses to pairs with a split of 2 were faster than responses to pairs with a split of 1 [816 vs. 959msec; $t(286)$ = 5.0, $P < .0001$]. Furthermore, latencies tended to increase with the *min* of the pair (r = .238, $P < .001$).

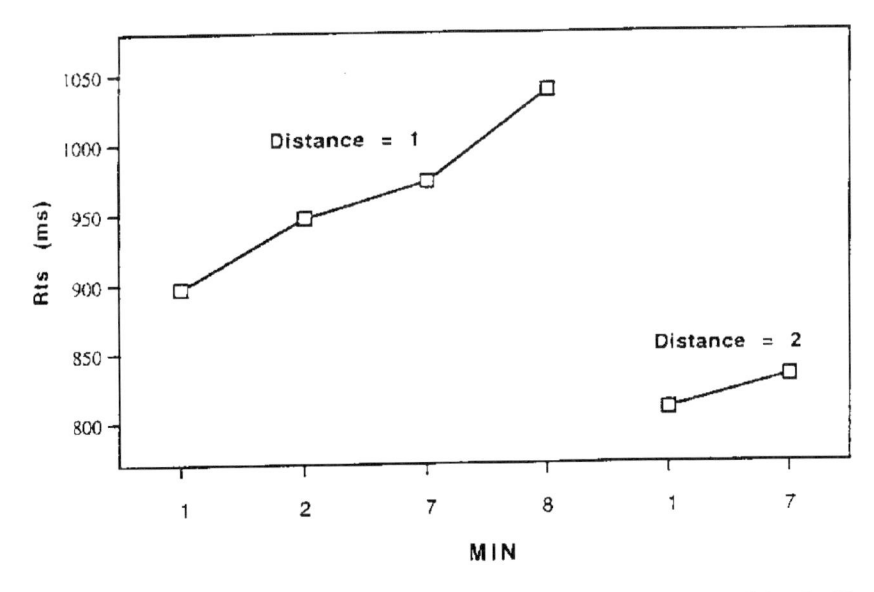

Fig. 8. GV's distribution of correct response latencies in a Magnitude Judgement Task with digit pairs (Banks et al., 1976). *Distance* = numerical difference between the digits (1 vs. 2); *Min* = value of the smaller digit of the pair.

Magnitude Judgements (Number Words)

The preceding tasks have documented that GV can access magnitude knowledge from visually presented Arabic numerals. Can she also access quantity information from visually presented number words or is access to this information only possible from Arabic numerals? To answer this question, GV was asked to determine whether a single number word was larger/smaller than 5 (not presented). The same procedure as that used for Arabic numerals was used in this task. GV successfully classified 45/64 (70%) digit words as larger/smaller than the standard 5—a far lower level of performance than the nearly perfect scores with Arabic numerals. Her level of performance in this task is not different from the word/picture matching task reported earlier (64% correct). In effect, having only to determine whether a word satisfies known conditions (either the pictured object or the number standard) helps her to focus on peculiar features of the word (e.g. length, particular characters) that may have been sufficient in some instances to produce a correct answer. However, the benefit of these cues was quite limited—GV was correct on only 70% of the trials, in a task where chance level was 0.5^3.

---

[3] That expectations about the material included in the list dramatically affected GV's responses is demonstrated by her performance in another task—the Letter/Digit Naming Task. With lists containing only letters or only digits, all the errors were "within-category" substitutions—letter → "letter" (e.g. *T* → "h") or digit → "digit" (e.g. *8* → "2"). In contrast, with triplets formed by various combinations of letters and digits (e.g. *1AL, F3H, STG*, etc.), "between-category" errors were produced not infrequently (e.g. *1AL* → "kp3").

Discussion

The results reported in this section unequivocally demonstrate that GV can access quantity information from Arabic numerals. She could accurately and rapidly compare Arabic numerals on the basis of magnitude. And when her response latencies were analysed, these replicated various effects related to quantity processing documented with neurologically intact subjects (*distance* effect, e.g. Banks et al., 1976; Moyer & Landauer, 1967; *min* effect, e.g. Parkman, 1971; see Figs. 7 and 8). However, the availability of numerical quantity was restricted to Arabic numerals[4]. When numerals were presented as visual words, access to quantity information was quite limited. The implications of the results in the "quantity" tasks will be examined in the General Discussion. Here we wish to draw attention to GV's extraordinarily efficient performance in the magnitude judgement tasks with Arabic numerals. GV's essentially "normal" performance in these tasks renders quite implausible an explanation of her difficulties in processing graphemes solely in terms of a deficit in the perceptual analysis of alphanumeric characters. A more plausible explanation is one that points a deficit at some level beyond *perceptual* analysis, more specifically at a point where digits are distinguished from letters.

GENERAL DISCUSSION

We have reported the detailed investigation of a brain-damaged patient (GV) who presents with a deficit in processing pictures, colours, and visually presented letters, words, and numerals. On the basis of evidence converging from multiple sources we can determine precisely the type of functional damage that underlies GV's reading impairment.

Several lines of evidence argue against a low-level perceptual deficit as the cause of GV's reading impairment. GV performed normally in a series of very demanding visual discrimination tests (see Table 4) and in tasks in which accurate performance depends on the ability to process adequately the form of visual letters. Thus, she correctly differentiated between real letters and pseudoletters (see Fig. 3) and she performed normally in a very demanding task involving the discrimination of canonical letter orientations (see Fig. 4). Furthermore, we reported evidence that her visual processing is sufficiently intact to support the recognition of digits—stimuli not unlike letters—since she performed normally in magnitude judgement tasks (see Figs. 7 and 8).

In contrast to her apparently intact ability to process visual shapes, GV showed severe difficulties in accessing a letter's grapheme representation. She was only 62% correct in

---

[4] And even here there were various restrictions. Thus, for example, GV failed in making parity judgements (i.e. "is 2 even or odd?"). Furthermore, in naming Arabic digits, no preference was found for producing errors with a numerical value close to that of the target (e.g. 3 → "4", 7 → "9"; see Fig 2.).

deciding whether two letters of different shapes had the same or different names (e.g. A–a vs. A–B or A–b; see Fig. 5), and she performed extremely poorly in transcoding (writing) letters from upper to lower or lower to upper case (e.g. A → a, d → D; 37% correct).

Lexical processing results show that GV cannot access the orthographic representation of words from visual input, and consequently cannot access their meanings either. GV performed at chance level in a lexical decision task. The latter finding cannot be explained as the result of damage to knowledge of the orthographic structure of words accessed in normal reading. Her perfect ability to recognise orally spelled words (see Table 3) indicates that her knowledge of the orthographic structure of words is intact.

GV's access to semantics from print is also severely impaired. This conclusion is supported by two results: She performed at chance level in a semantic categorisation task with visual words and she failed to show a semantic priming effect when words were used as primes (see Fig. 6). These findings with words contrast sharply with those obtained with pictures and Arabic numerals. With the latter stimuli, evidence was found of partial activation of semantic information. Thus, for example, GV successfully sorted visual objects by function and she could access quantity information for digits. These contrasting results suggest that a mechanism essential for the processing of words but not for the processing of pictures and Arabic numerals is selectively damaged (or disconnected from visual input).

Taken together, the results reported here allow a precise characterisation of the causes of GV's reading impairment: It originates from a failure to access graphemic representations from the normally computed visual letter shapes. As a consequence of this deficit, GV cannot access her intact lexical orthographic representations and associated meanings. The remainder of this section of the paper explores several implications that follow from this conclusion.

The Early Stages of the Reading Process

The analysis of GV's reading impairment revealed a series of clear-cut dissociations that jointly constrain the architecture of the early stages of visual word recognition. Of particular relevance here is the striking dissociation between her ability to access the visual shape information for letters but not their graphemic identity. This dissociation together with the observation that she is totally unable to retrieve lexical or semantic information from visual words has rather strong implications for models of reading.

A straightforward implication that follows from the dissociation in processing letter shapes versus grapheme identity is that there are two distinct types of letter representations computed in reading: First, a representation that encodes letters as specific visual objects with particular canonical orientations (e.g. A, *A*, a, *a*, etc,), and, second, a representation that specificies abstract letter identities independently of their specific visual forms (e.g. the grapheme "a"). This conclusion is far from unexpected: There is a plenitude of results in cognitive psychology suggesting a distinction of this sort (e.g. Adams, 1979). The relevance of our results, however, resides in their implications for the level of visual letter representation that mediates the access to lexical orthographic representations, an issue that remains controversial (Coltheart, 1981; Henderson, 1982; Mayall, Humphreys, & Olson, 1997). The fact that our patient cannot access graphemic representations and the fact that her access to the orthographic content of visually presented words seems to be almost completely abolished invite the conclusion that grapheme identities mediate access to the orthographic lexicon.

Several types of results have been interpreted as demonstrating that the orthographic content of a word is stored in an abstract format. These results include, among others, evidence of word superiority effects with alternating case stimuli (e.g. tAbLe; see, for example, Adams, 1979; Besner, 1989; McClelland, 1976; but see also Coltheart & Freeman, 1974), and the finding that a change of letter case during a saccadic eye movement does not interfere with word reading (Rayner, McConkie, & Zola, 1980). Our results further strengthen this conclusion, in that they suggest that the normal computation of a graphemic representation is necessary for access to the orthographic lexicon and subsequent access to the semantic system[5]. This hypothesis is shown in Fig. 9. Information concerning the shape of letters or the contour of the word does not seem to play a crucial role in word identification, as evidenced by the fact that, despite the availability of this type of information to our patient, she failed to identify words. This conclusion is consistent with previous results demonstrating that aspects related to the shape of letters or words are of extremely limited importance (if any) in word identification (e.g. Henderson & Chard, 1976; Mayall et al., 1997; Monk & Hulme, 1983; Underwood & Bargh, 1982; see Henderson, 1982 for review). They are also consistent with results from neglect dyslexia, where it has been shown that the spatially specific deficit can affect either a level of processing where letter shape information is represented or a level where graphemic information is represented (e.g. Caramazza & Hillis, 1990; Hillis & Caramazza, 1995b).

Reading Processing in the Right Hemisphere?

The hypothesis that the right hemisphere has considerable reading capabilities continues to be a contentious issue in cognitive neuropsychology (see, e.g. Baynes, 1990; Coltheart, 1980; Gazzaniga, 1989; Patterson & Besner, 1984; Zaidel & Peters, 1981). Coslett and Saffran (1989a, 1989b; 1994; see also Saffran & Coslett, this issue) have proposed that the right hemi-

---

[5] Findings that are apparently in conflict with this proposal were reported by Howard (1987). His patient could not indicate whether letters of different shape have the same name but could read a few words (about 20% of the stimuli). However, this level of performance may simply reflect the guessing level supported by limited access to orthographic information.

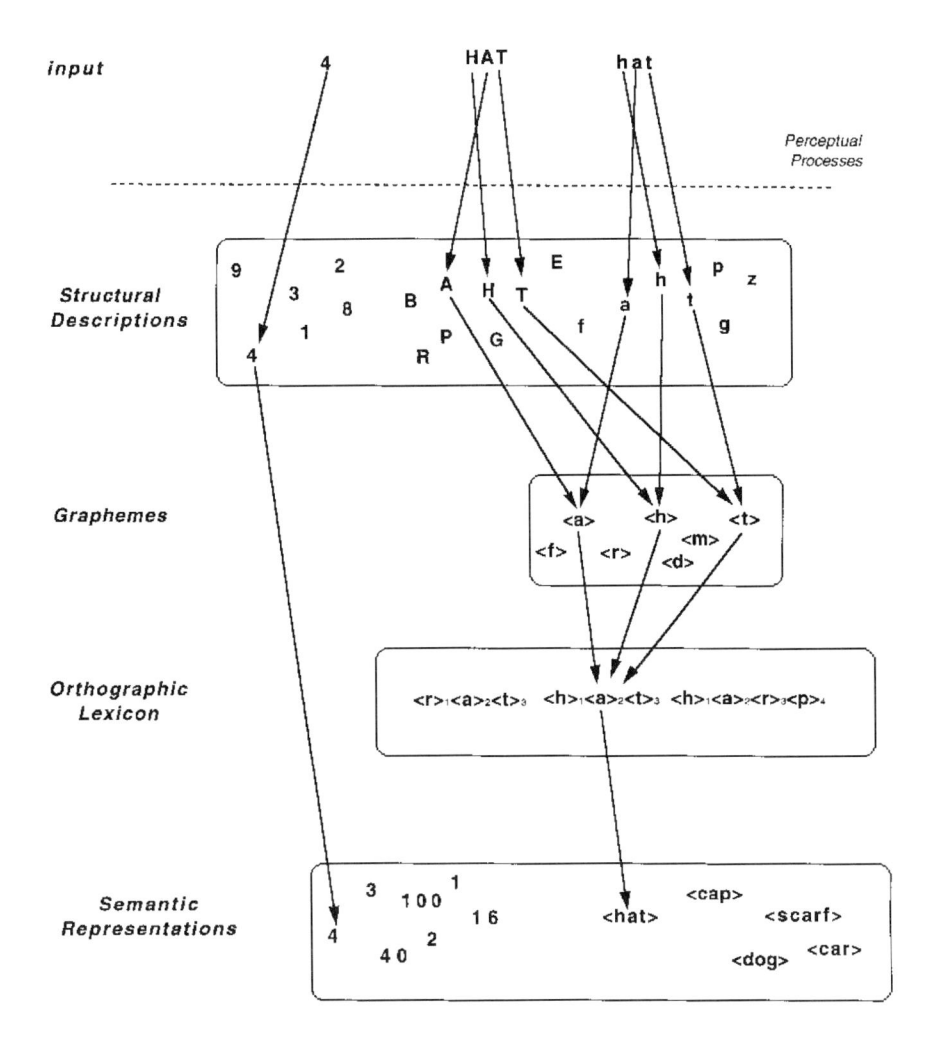

Fig. 9. Schematic representation of the access to orthographic and semantic information with words and of the access to semantic (quantity) information with numerals.

sphere is able to process all types of information about visually presented concrete words with the exception of their phonological and morphological features. In other words, the right hemisphere can process graphemic representations and access lexical orthographic representations and their associated meanings. They base their claim on results that suggest that pure alexic subjects have access to the orthographic and semantic content of words despite their poor performance in explicit word recognition and oral reading. These results are obtained in conditions in which reading processing in the left hemisphere is supposedly prevented through rapid presentation of stimuli and rapid responding. Under these conditions, patients with pure alexia exhibit surprising reading abilities. Thus, for example, patient CB (Coslett & Saffran, 1989b) although completely unable to pronounce any written word, performed at above-chance level in lexical decision and semantic categorisation tasks with briefly presented visual words. Coslett and Saffran attributed these residual abilities to the functioning of reading processes in the right hemisphere. This conclusion rests principally on the argument that pure alexia is (usually) caused by damage to left occipital areas and the corpus callosum, thereby effectively isolating the left-hemisphere language areas from visual input[6]. As a consequence, the observed behaviour in these patients reflects *directly* (in appropriate test conditions) the functioning of the right hemisphere. This account provides a straightforward explanation of the reading impairment found in pure alexic patients[7].

However, the data from GV appear to be problematic for the proposal of extensive right-hemisphere reading abilities. Despite the fact that in many respects our case is similar to those presented by Coslett and Saffran, we failed to observe any spared reading ability in our patient (see also Chialant & Caramazza, this issue). GV's reading deficit was the result of left posterior damage that extended to the corpus callosum. With visually presented objects, GV demonstrated intact access to structural/visual knowledge and she could discriminate between semantic categories. Her pattern of performance in tasks with visual objects conforms to that of other cases having the same type of brain damage that has been reported in the literature (e.g. Coslett & Saffran, 1989b, 1992; De Renzi & Saetti, 1997; Hillis & Caramazza, 1995a; Iorio et al., 1992; Manning & Campbell, 1992; Riddoch & Humphreys, 1987b). Furthermore, her abilities to access quantity information from Arabic digits parallel those described by Cohen and Dehaene (1995) in patients with left occipital and callosal lesions (resulting in pure alexia). Yet, our patient did not show the same reading abilities as the cases documented by Coslett

---

[6] Bowers, Bub, and Arguin (1996) reported the interesting case of a letter-by-letter reader who showed a similar pattern of performance with words briefly displayed to the right hemisphere in familiar and unfamiliar (e.g. cHaIr) format. This finding suggests that the patient can access word orthographic knowledge. Whether the patient retrieves orthographic information stored in the right as opposed to the left hemisphere, however, is not resolved by the results reported in the paper.

[7] Although Coslett and Saffran's account of the reading mechanisms in the right hemisphere is the most explicit formulation yet, there are several important aspects that are still unspecified. For example, what is the function of the right-hemisphere reading mechanism in the normal system? How does the information computed in the right hemisphere interact with that computed in the left hemisphere? Is the information in the right hemisphere merely a pale duplicate of that in the left? And what about the assumption that only homologous regions in the two hemispheres are connected? Does this assumption mean that the connections between the two hemispheres involve only "identical" sorts of information, such that activation of particular representations in the right hemisphere activate their corresponding forms in the left hemisphere (the duplicate activation hypothesis)? What is the function of these connections in the normal reading system?

and Saffran. In fact, GV performed at chance level in a lexical decision task and a word categorisation task (animal vs. artefact), and she did not show priming effects for words. These results were obtained despite the fact that, as described earlier, we followed exactly the same procedures as those used by Coslett and Saffran. Thus, when all the behavioural facts of our case are taken into account, one is forced to conclude that the reading deficit in our patient is left unexplained by the proposal of reading functions in the right hemisphere.

One way to reconcile the conflicting data of GV and the right-hemisphere hypothesis is to postulate that the reading capabilities of the right hemisphere vary extensively across individuals. At one end of the continuum are subjects like GV, who do not have extensive reading mechanisms in the right hemisphere; at the other end there are subjects like those reported by Coslett and Saffran who appear to have considerable right-hemisphere reading capabilities, and there are individuals with varying intermediate capabilities between these two extremes.

Another way to reconcile the data of GV and the cases reported by Coslett and Saffran is to assume an additional lesion in GV's right hemisphere that selectively damages the mechanisms supporting grapheme recognition in that hemisphere. And one might propose that these structures reside in the parietal area, which appeared to be damaged in our patient. We have no direct evidence that would exclude this possibility. Nonetheless, there are several arguments that could be offered against this proposal. First, the patient did not manifest any of the clinical signs associated with parietal damage, including visual neglect or other forms of visuospatial impairments. Her intact performance in the very demanding Cooper and Shepard's Rotation Task testifies to the preservation of visuospatial functions. Second, a patient with a pattern of reading performance very similar to that of GV but with no apparent right-side lesion has recently been reported by Chanoine, Teixeira Ferreira, Demonet, Nespolous, and Poncet (submitted; for a description of this case see also Teixeira Ferreira, Giusiano, Ceccaldi, & Poncet, 1997). In this patient (CN), neurobiological examination (MRI) revealed only a left occipito-temporal lesion, extending subcortically to the callosal fibres. CN was completely unable to read. Furthermore, although able to discriminate between letters and pseudoletters, he was unable to categorise letters shown in different fonts as representing the same grapheme. In essence, CN and GV showed identical functional deficits: failure to access the graphemic representations of normally computer letter shapes. However, and of particular relevance in this context, CN's deficit cannot be ascribed to right parietal damage, Thus, this case provides support for the claim that our patient's inability to compute graphemic representations is not caused by a right parietal lesion.

Another problematic case for the right-hemisphere hypothesis has been reported by Chialant and Caramazza (this issue). This patient (MJ) presented with classical symptoms of the letter-by-letter form of pure alexia. Unlike the several cases reported by Coslett and Saffran (1994), MJ did not show a semantic priming effect for words she could not recognise explicitly when sophisticated guessing was taken into account. Crucially for our argument here, MJ has damage only to the left hemisphere (occipital and posterior subcortical), just like CN. Thus, her failure to access meaning implicitly from print cannot be ascribed to damage to the right hemisphere.

The results of our investigation suggest that two types of letter representations are computed in the course of visual word recognition: One specifies the shape and orientation of letter forms, and the other specifies their graphemic identity. The fact that GV failed to access graphemic representations despite normal access to letter forms has immediate implications for the anatomical organisation of the early stages of reading processes. Namely, these facts imply that only representations about the shape and orientation of letters are encoded in the right hemisphere; graphemic representations of letters are stored in the left hemisphere[8]. However, as noted above, this may not generalise to all individuals.

Words, Numbers, and Pictures

In our discussion thus far, we have focused on reading to the complete exclusion of other aspects of our patient's cognitive deficits. Thus, we have not discussed GV's ability to extract (partial) meaning from pictures and numbers despite her very poor performance in naming these stimuli. We showed that GV can recognise whether two visually presented objects belong to the same semantic category, but she fails when fine-grained semantic distinctions are required. Particularly striking is GV's performance with numerals: Although she is able to indicate in a fraction of a second whether an Arabic digit is larger or smaller than 5, she could not perform this task with number words even after laborious attempts (for a similar dissociation, see Cohen & Dehaene, 1995). What are we to make of these dissociations? Why is it that GV can access (some) meaning from pictures and Arabic numerals but not words?

We have argued that the reason for GV's inability to access meaning from written words

---

[8] Results consistent with this hypothesis are found in studies with neurologically intact subjects (e.g. Cohen, 1972; Geffen, Bradshaw, & Nettleton, 1972; Ledlow, Swanson, & Kinsbourne, 1978; Umiltà, Sava, & Salmaso, 1980; but see Boles, 1981). Using a divided-visual-field technique, it has been shown that subjects recognise faster that two letters are physically identical (e.g. A–A) when the stimuli are presented to the right hemisphere (left visual hemifield). In contrast, the decision that two letter forms are the same grapheme (e.g. A–a) are made faster when the stimuli are shown to the left hemisphere (right visual hemifield). The advantage of the right hemisphere with letter shape judgement has been interpreted as implying that this type of information is represented (or processed more efficiently) in that hemisphere (on this point see also Bryden & Allard, 1976; Marsolek, Squire, Kosslyn, & Lulenski, 1994). The disadvantage of the right hemisphere for letter name judgements has been interpreted as indicating that information in the left hemisphere is needed for this type of response.

However, there are also results that are potentially problematic for the claim that graphemic information is represented only in the left hemisphere. Eviatar and Zaidel (1994) report three split-brain patients who showed some residual abilities to recognise whether two letters (e.g. A–a) were the same grapheme when they were presented to the right hemisphere.

is because of the hemispheric disconnection between the intact graphemic representations in her left hemisphere and the intact letter form information in her right hemisphere. Since lexical access is driven by graphemic representations, the disconnection between the letter shape and the graphemic levels of representation effectively blocks this process, leading to severe difficulties in recognising words and accessing their meaning. This claim rests on the strong assumption that access to the meaning of words is mediated by processes in the left hemisphere. Is this assumption equally plausible for the case of pictures and Arabic numerals?

If we were to answer affirmatively, we would have to explain why it is that the structural representations of pictures and Arabic numerals in the right hemisphere can access their semantic representations in the left hemisphere, albeit not perfectly. One possible explanation is that the paths of interhemispheric transfer for these representations are different from that for letters. And on the further assumption that undamaged interhemispheric transfer is "noisy" (as opposed to being severely damaged as we have argued for the transmission path for letter), we can explain why there is partial access to meaning for pictures and Arabic numerals, but not words. However, this explanation does not provide a principled account for the dissociation between access to quantity information and other number facts such as parity information.

However, results conflicting with those of Eviatar and Zaidel are reported by Reuter-Lorenz and Baynes (1992). With stimuli presented to the right hemisphere, their split-brain patient showed no evidence of cross-case priming. Priming was obtained only when the stimuli were shown to the left hemisphere. Unfortunately, it is unclear whether such discrepant findings are the result of the different methods used to assess the availability of graphemic information or whether they might reflect critical discrepancies in hemispheric organisation among these patients. Thus, it is premature to draw firm conclusions on the basis of these data.

Other potentially problematic results come from the study of a right-handed teenager (NI) who had undergone complete left-hemispherectomy (Patterson, Vargha-Khadem, & Polkey, 1989). Until age 13, when the subject developed an intractable epilepsy, language and reading skills appeared to be normal. Postoperative testing revealed spared reading abilities, which included preserved cross-case letter matching, and above-chance accuracy in lexical decision and word categorisation tasks. Far more remarkable in this case is the appearance of speech after the operation. NI's speech was fairly rich as evinced by, for instance, her good performance in naming pictured actions (76% correct). NI's extensive postoperative speech production ability suggests "abnormally" lateralised language.

Apparently more promising are the analyses of the reading performance of adult subjects with selective damage of the posterior portion of the corpus callosum. For these cases, there is no reason to suspect abnormally developed hemispheric organisation. However, conflicting patterns of reading performance have been reported for these patients. The patient described by Michel, Hénaff, and Intriligator (1996), for instance, performed fairly accurately in a lexical decision task, (95%) with written words presented to the right hemisphere. With the same technique of stimulus presentation, however, the patient reported by Cohen and Dehaene (1996) performed at chance level in a word semantic categorisation task. The reasons of these discrepancies are unclear: Do they reflect different forms of hemispheric organisation? Or do they reflect variation in the information that can be transferred to the left hemisphere?

The other possibility is that the right hemisphere is capable of representing number magnitude information and various conceptual features of objects but not complete semantic representations for either class of objects. With these assumptions we can explain why GV is able to categorise objects and make magnitude judgements with Arabic numerals (see also Dehaene, 1995, 1996). To explain why GV fails to name Arabic numerals and pictures correctly, we further need to assume that semantic information in the right hemisphere is insufficient to drive the lexical system in the left hemisphere without error, either because the semantic information is incomplete or because it becomes degraded during interhemispheric transfer. However, this story too is incomplete in important ways. Thus, for example, what relationships are there between the "semantics" of objects and numbers in the two hemispheres? What connection is there between these "semantics" and the semantics of words? Are they the same kind of things? Unfortunately, we do not have answers to these and the many other questions that one could raise about the distribution of functions across the two hemispheres. However, the study of patients such as GV may eventually contribute to a clarification of these important issues.

REFERENCES

Adams, M.J. (1979). Models of word recognition. *Cognitive Neuropsychology, 11*, 133–176.

Arguin, M., & Bub, D.N. (1993). Single-character processing in a case of pure alexia. *Neuropsychologia, 31*, 435–458.

Banks, W.P., Fujii, M., & Kayra-Stuart, F. (1976). Semantic congruency effects in comparative judgements of magnitude of digits. *Journal of Experimental Psychology: Human Perception and Performance, 2*, 435–477.

Baynes, K. (1990). Language and reading in the right hemisphere: Highways or byways of the brain? *Journal of Cognitive Neuroscience, 2*, 159–179.

Behrmann, M., Plaut, D.C., & Nelson, J. (this issue). A literature review and new data supporting an interactive account of letter-by-letter reading. *Cognitive Neuropsychology, 15*(1/2).

Benson, D.F. (1985). Alexia. In G.W. Bruyn, H.K. Klawans, & P.J. Vinken (Eds.), *Handbook of clinical neurology*. (pp. 433–455). New York: Elsevier.

Besner, D. (1989). On the role of outline shape and word-specific visual pattern in the identification of function words: NONE. *Quarterly Journal of Experimental Psychology, 41A*, 91–105.

Black, S.E., & Behrmann, M. (1994). Localization in alexia. In A. Kertesz (Ed.), *Localization and neuroimaging in neuropsychology* (pp. 331–376). San Diego, CA: Academic Press.

Boles, D.B. (1981). Variability in letter-matching asymmetry. Perception & Psychophysics, 29, 285–288.

Boles, D.B., & Clifford, J.E. (1989). An upper- and lower-case alphabetic similarity matrix, with derived generation similarity values. *Behavior Research Methods, Instruments, and Computers, 21*, 579–586.

Bowers, J.S., Bub, D.N., & Arguin, M. (1996). A characterisation of the word superiority effect in a case of letter-by-letter surface alexia. *Cognitive Neuropsychology, 13*, 415–441.

Bryden, M.P., & Allard, F. (1976). Visual hemifield differences depends on typeface. *Brain and Language, 3*, 191–200.

Bub, D.N., & Arguin, M. (1995). Visual word activation in pure alexia. *Brain and Language, 49*, 77–103.

Buckley, P.B., & Gillman, C.B. (1974). Comparison of digits and dot patterns. *Journal of Experimental Psychology, 103,* 1131–1136.

Caramazza, A., & Hillis, A.E. (1990). Levels of representations, co-ordinate frames, and unilateral neglect. *Cognitive Neuropsychology, 7,* 391–445.

Chanoine, V., Teixeira Ferreira, C., Demonet, J.F., Nespoulous, J.L., & Poncet, M. (submitted). *Optic aphasia and alexia without agraphia: Two variants of visual associative agnosia? A case study.* Manuscript submitted for publication.

Chialant, D., & Caramazza, A. (this issue). Perceptual and lexical factors in a case of letter-by-letter reading. *Cognitive Neuropsychology, 15*(1/2).

Cipolotti, L., & Butterworth, B. (1995). Toward a multiroute model of number processing: Impaired number transcoding with preserved calculation skills. *Journal of Experimental Psychology: General, 124,* 375–390.

Cipolotti, L., Warrington, E.K., & Butterworth, B. (1995). Selective impairment in manipulating Arabic numerals. *Cortex, 31,* 73–86.

Cohen, G. (1972). Hemispheric differences in a letter classification task. *Perception & Psychophysics, 11,* 139–142.

Cohen, L., & Dehaene, S. (1995). Number processing in pure alexia: The effect of hemispheric asymmetries and task demands. *Neurocase, 1,* 121–137.

Cohen, L., & Dehaene, S. (1996). Cerebral networks for number procesesing: Evidence from a case of posterior callosal lesion. *Neurocase, 2,* 155–174.

Cohen, L., Dehaene, S., & Verstichel, P. (1994). Number words and number nonwords: A case of deep dyslexia extending to Arabic numerals. *Brain, 117,* 267–279.

Coltheart, M. (1980). Deep dyslexia: A right hemispheric hypothesis. In M. Coltheart, K. Patterson, & J.C. Marshall (Eds.), *Deep dyslexia* (pp. 326–380). London: Routledge & Kegan Paul.

Coltheart, M. (1981). Disorders of reading and their implications for models of normal reading. *Visible Language, 15,* 245–286.

Coltheart, M., & Freeman, R. (1974). Case alternation impairs word identification. *Bulletin of the Psychonomic Society, 3,* 102–104.

Cooper, L.A., & Shepard, R.N. (1973). Chronometric studies of the rotation of mental images. In W.G. Chase (Ed.), *Visual information processing* (pp. 75–175). New York: Academic Press.

Coslett, H.B., & Monsul, N. (1994). Reading with the right hemisphere: Evidence from transcranial magnetic stimulation. *Brain and Language, 46,* 198–211.

Coslett, H.B., & Saffran, E.M. (1989a). Evidence for preserved reading in "pure alexia." *Brain, 112,* 327–359.

Coslett, H.B., & Saffran, E.M. (1989b). Preserved object recognition and reading comprehension in optic aphasia. *Brain, 112,* 1091–1110.

Coslett, H.B., & Saffran, E.M. (1992). Optic aphasia and the right hemisphere: A replication and extension. *Brain and Language, 43,* 148–161.

Coslett, H.B., & Saffran, E.M. (1994). Mechanisms of implicit reading in alexia. In M.J. Farah & G. Ratcliff (Eds.), *The Neuropsychology of high-level vision* (pp. 299–330). Hillsdale, NJ: Lawrence Erlbaum Associates Inc.

Damasio, A.R., & Damasio, H. (1983). The anatomic basis of pure alexia. *Neurology, 33,* 1573–1583.

Dehaene, S. (1995). Towards an anatomical and functional model of number processing. *Mathematical Cognition, 1,* 83–120.

Dehaene, S. (1996). The organization of brain activation in number comparison: Event related potentials and the additive-factors method. *Journal of Cognitive Neuroscience, 8,* 47–68.

Dehaene, S., & Cohen, L. (1996). Two mental calculation systems: A case study of severe acalculia with preserved approximation. *Neuropsychologia, 29,* 1045–1074.

Déjèrine, J. (1892). Contribution a l'etude anatomo-pathologique et clinique des differentes varietes de cicite verbale. [Contributions to the anatomic, pathological, and clinical investigation of different varieties of word blindness]. *Memoires de la Societe de Biologie, 4,* 61–90.

De Renzi, E., & Saetti, M.C. (1997). Associative agnosia and optic aphasia: Qualitative or quantitative difference? *Cortex, 33*, 115–130.

Eviatar, Z., & Zaidel, E. (1994). Letter matching within and between the disconnected hemispheres. *Brain and Cognition, 25*, 128–137.

Farah, M.J., & Wallace, M.A. (1991). Pure alexia as a visual impairment: A reconsideration. *Cognitive Neuropsychology, 8*, 313–334.

Friedman, R.B., & Alexander, M.P. (1984). Pictures, images, and pure alexia: A case study. *Cognitive Neuropsychology, 1*, 9–23.

Gazzaniga, M.S. (1989). Organization of the human brain. *Science, 245*, 947–952.

Geffen, G., Bradshaw, J.L., & Nettleton, N.C. (1972). Hemispheric asymmetry: Verbal and spatial encoding of visual stimuli. *Journal of Experimental Psychology, 95*, 25–31.

Geschwind, N., & Fusillo, M. (1966). Colour-naming defects in association with alexia. *Archives of Neurology, 15*, 137–146.

Goodman, R., & Caramazza, A. (1986). *The Johns Hopkins Dyslexia Battery*. Baltimore, MD: The Johns Hopkins University.

Henderson, L. (1982). *Orthography and word recognition in reading*. London: Academic Press.

Henderson, L., & Chard, M.J. (1976). On the nature of facilitation of visual comparisons by lexical membership. *Bulletin of the Psychonomic Society, 7*, 432–434.

Hillis, A.E., & Caramazza, A. (1995a). Cognitive and neural mechanisms underlying visual and semantic processing: Implications from "optica aphasia." *Journal of Cognitive Neuroscience, 7*, 457–478.

Hillis, A.E., & Caramazza, A. (1995b). A framework for interpreting distinct patterns of hemispatial neglect. *Neurocase, 1*, 189–207.

Hinrichs, J.V., Yurko, D.S., & Hu, J.M. (1981). Two-digit number comparison: Use of place information. *Journal of Experimental Psychology: Human Perception and Performance, 7*, 890–901.

Howard, D. (1987). Reading without letters? In M. Coltheart, G. Sartori, & R. Job (Eds.), *The cognitive neuropsychology of language* (pp. 27–58). Hove, UK: Lawrence Erlbaum Associates Ltd.

Howard, D. (1991). Letter-by-letter readers: Evidence for parallel processing. In D. Besner & G.W. Humphreys (Eds.), *Basic processes in reading: Visual word recognition*. Hillsdale, NJ: Lawrence Erlbaum Associates Inc.

Iorio, L., Falanga, A., Fragrassi, N.A., & Grossi, D. (1992). Visual associative agnosia and optic aphasia. A single case study and a review of the syndromes. *Cortex, 28*, 23–37.

Kay, J., & Hanley, R. (1991). Simultaneous form perception and serial letter recognition in a case of letter-by-letter reading. *Cognitive Neuropsychology, 8*, 249–273.

Kinsbourne, M., & Warrington, E.K. (1962). A disorder of simultaneous form perception. *Brain, 85*, 461–486.

Ledlow, A., Swanson, J.M., & Kinsbourne, M. (1978). Reaction times and evoked potentials as indicators of hemispheric differences for laterally presented name and physical matches. *Journal of Experimental Psychology: Human Perception and Performance, 4*, 440–454.

Levine, D.M., & Calvanio, R. (1992). A study of the visual defect in verbal alexia-simultagnosia. *Brain, 101*, 65–81.

Manning, L., & Campbell, R. (1992). Optic aphasia with spared action naming: A description and possible loci of impairment. *Neuropsychologia, 30*, 587–592.

Marsolek, C.J., Squire, L.R., Kosslyn, S.M., & Lulenski, M.E. (1994). Form-specific explicit and implicit memory in the right cerebral hemisphere. *Neuropsychology, 8*, 588–597.

Mayal, K., Humphreys, G.W., & Olson, A. (1997). Disruption to word or letter processing? The origins of case-mixing effects. *Journal of Experimental Psychology: Learning, Memory, and Cognition, 23*, 1275–1286.

McClelland, J.L. (1976). Preliminary letter identification in the perception of words and nonwords. *Journal of Experimental Psychology: Human Learning and Memory, 2*, 80–91.

McNeil, J.E., & Warrington, E.K. (1994). A dissociation between addition and subtraction with written calculation. *Neuropsychologia, 32*, 717–728.

Michel, F., Hénaff, M., & Intriligator, J. (1996). Two different readers in the same brain after a posterior callosal lesion. *NeuroReport, 7*, 786–788.

Monk, A.F., & Hulme, C. (1983). Errors in proofreading: Evidence for the use of word shape in word recognition. *Memory and Cognition, 11*, 16–23.

Montant, M., Nazir, T.A., & Poncet, M. (this issue). Pure alexia and the viewing position effect in printed words. *Cognitive Neuropsychology, 15*(1/2).

Moyer, R.S., & Landauer, T.K. (1967). Time required for judgements of numerical inequality. *Nature, 215*, 1519–1520.

Parkman, J.M. (1971). Temporal aspects of digit and letter inequality judgements. *Journal of Experimental Psychology, 91*, 191–205.

Patterson, K., & Besner, D. (1984). Is the right hemisphere literate? *Cognitive Neuropsychology, 4*, 315–341.

Patterson, K., & Kay, J. (1982). Letter-by-letter reading: Psychological descriptions of a neurological syndrome. *Quarterly Journal of Experimental Psychology, 34A*, 411–441.

Patterson, K., Vargha-Khadem, F., & Polkey, C.E. (1989). Reading with one hemisphere. *Brain, 112*, 39–63.

Perri, R., Bartolomeo, P., & Silveri, A.M. (1996). Letter dyslexia in a letter-by-letter reader. *Brain and Language, 53*, 390–407.

Posner, M.I., & Mitchell, R.F. (1967). Chronometric analysis of classification. *Psychological Review, 74*, 391–409.

Price, C.J., & Humphreys, G.W. (1992). Letter-by-letter reading? Functional deficits and compensatory strategies. *Cognitive Neuropsychology, 9*, 427–457.

Rapp, B.C., & Caramazza, A. (1991). Spatially determined deficits in letter and word processing. *Cognitive Neuropsychology, 8*, 275–311.

Rayner, L., McConkie, G.W., & Zola, D. (1980). Integrating information across eye movements. *Cognitive Psychology, 12*, 206–226.

Reuter-Lorenz, P.A., & Baynes, K. (1992). Modes of lexical access in the callosotomized brain. *Journal of Cognitive Neuroscience, 4*, 155–164.

Reuter-Lorenz, P.A., & Brunn, J.L. (1990). A prelexical basis for letter-by-letter reading: A case study. *Cognitive Neuropsychology, 7*, 1–20.

Riddoch, M.J., & Humphreys, G.W. (1987a). Picture naming. In G.W. Humphreys & M.J. Riddoch (Eds.), *Visual object processing: A cognitive neuropsychological approach* (pp. 107–143). Hillsdale, NJ: Lawrence Erlbaum Associates Inc.

Riddoch, M.J., & Humphreys, G.W. (1987b). Visual object processing in optic aphasia: A case of semantic access agnosia. *Cognitive Neuropsychology, 4*, 131–185.

Riddoch, M.J., & Humphreys, G.W. (1993). BORB Birmingham object recognition battery. Hove, UK: Lawrence Erlbaum Associates Ltd.

Saffran, E.M., Bogyo, L.C., Schwartz, M.F., & Marin, O.S.M. (1980). Does deep dyslexia reflect right-hemisphere reading? In M. Coltheart, K. Patterson, & J.C. Marshall (Eds.), *Deep dyslexia* (pp. 381–406). London: Routledge & Kegan Paul.

Saffran, E.M., & Coslett, H.B. (this issue). Implicit vs. letter-by-letter reading in pure alexia: A tale of two systems. *Cognitive Neuropsychology, 15*(1/2).

Sekuler, E.B., & Behrmann, M. (1996). Perceptual cues in pure alexia. *Cognitive Neuropsychology, 13*, 941–974.

Sekuler, R., Rubin, E., & Armstrong, R. (1971). Processing numerical information: A choice time analysis. *Journal of Experimental Psychology, 90*, 75–80.

Shallice, T., & Saffran, E. (1986). Lexical processing in the absence of explicit word identification: Evidence from a letter-by-letter reader. *Cognitive Neuropsychology, 3*, 429–458.

Snodgrass, J.G., & Vanderwart, M. (1980). A standardized set of 260 pictures: Norms for name agreement, image agreement, familiarity and visual complexity. *Journal of Experimental Psychology: Human Learning and Memory, 6*, 174–215.

Teixeira Ferreira, C., Giusiano, B., Ceccaldi, M., & Poncet, M. (1997). Optic aphasia: Evidence of the contribution of different neural systems to object and action naming. *Cortex, 33*, 499–513.

Umiltà, C., Sava, D., & Salmaso, D. (1980). Hemispheric asymmetries in a letter classification task with different typefaces. *Brain and Language, 9*, 171–181.

Underwood, G., & Bargh, K. (1982). Word shape, orthographic regularity, and contextual interactions in a reading task. *Cognition, 12*, 197–209.

Warrington, E.K., & Shallice, T. (1980). Word-form dyslexia. *Brain, 103*, 99–112.

Zaidel, E., & Peters, A.M. (1981). Phonological encoding and ideographic reading by the disconnected right hemisphere: Two case studies. *Brain and Language, 14*, 205–234.

Zaidel, E., & Schweiger, A. (1984). On wrong hypotheses about the right hemisphere: Commentary on K. Patterson and D. Besner, "Is the right hemisphere literate?" *Cognitive Neuropsychology, 4*, 351–364.

Subject Index

Affix sensitivity, 145, 147, 158
Aphasia, 159–160
Attractors, 16
Aural recognition, 178–179, 212–213

Brain lesion site & reading defect, 151–152

Cascaded processing, 13
Children
    viewing position effect, 127
    word length effect, 127
Colour naming, 210
Compensation strategies, 43
Concreteness, see Imageability
Connectionist models, 16–17, 39
Covert lexical activation, 55–57, 79–80
    implicit reading, 80–82
    left hemisphere, 68, 81
    phonological access, 71
    priming, 57, 58, 66, 67, 81
Covert processing, 5, 24
Covert reading, 54–56, 61, 80–82, 144–162, 188–195
Cross-case letter-matching, 144, 219

Definitions, naming objects from, 210
Degraded stimuli
    letter-by-letter readers, 143
    normal readers, 40–41
Disconnection theories, 11–12, 125–126, 142, 168, 198–199
Distributed representations, 16–17
Dorsal white matter, 151–152
Dual read-out model, 110

End-effect, 117
Experimental problems with letter-by-letter readers, 148

Fixation patterns, 14
Fixation position, see Viewing position effect
Frequency effects
    neighbourhood size, 75–79
    theories of, 29
    word length, 33, 34, 37
    word recognition accuracy, 25
    word recognition speed, 25, 29, 32–33, 75

Hemispherectomy, 157, 159

Identity priming, 192–194
Imageability
    naming speed, 33–34
    normal reading, 30
    reading aloud, 4
    semantic representations, 4
    word length, 33, 34, 37, 158
Implicit reading, 54–56, 61, 80–82, 144–162, 188–195
Interactive Activation Model, 3, 13, 126
Interactive processing, 13
Interactive theories
    letter-by-letter reading, 13–15, 17–18, 36
    neglect dyslexia, 15

Left hemisphere, covert lexical activation, 68, 81
Left-hemispherectomy, 157, 159
Lesion site & reading defect, 151–152
Letter-by-letter reading
    accompanying language deficits, 8
    compensation strategies, 43
    severity of, 42
    variability in, 41–43
Letter-by-letter reading models, 38–40
Letter-by-letter reading speed & stimulus quality, 143
Letter-by-letter reading theories, 54, 94–95, 168–170, 195–199, 204–205
    central vs peripheral, 8, 11–12
    disconnection, 11–12, 125–126, 142, 168, 198–199
    interactive, 13–15, 17–18, 36
    right hemisphere, 4, 12–13, 44–45, 61, 151, 170, 198–199, 204–205, 229–234
Letter matching, 144, 219
Letter position coding, 123–125, 183–185
Letter processing in letter-by-letter readers, 9–10, 19, 22–23, 143–144, 197
Letter recognition accuracy, 10, 143
    exposure time, 27
    position in letter string, 120–121, 123–125, 183–185
    upper vs lower case asymmetries, 210–211
    visual field asymmetries, 115, 117
Letter recognition speed, 19, 22
Letter representation when reading, 227–228
Letter search, 183–185
Letter transcoding, 219
Lexical decisions
    affix sensitivity, 145, 147, 158
    brief presentations, 10–11, 144–145
    frequency, 25
    words vs nonwords, 175–177, 220
    words vs pseudowords, 111–114
Line orientation detection, 185–186

Magnitude judgements, 157, 223–226, 232–234

Neglect dyslexia
    modelling, 15, 17
    orthographic neighbours, 75
    severity of, 42
    spatial deficits, 228

SUBJECT INDEX

Neighbourhood size, 72–79
Network models, 16–17, 39
Nonword reading, 3
Nonword vs word decisions, 175–177, 220
Number naming, 211–212
Number processing, 157, 212, 218, 223–226, 232–234

Object/picture naming, 143, 152, 154, 209, 232–234
    brief exposures, 169
    different perspectives, 214
    exposure time, 179–180
    visual complexity, 27, 29
Oral spelling, 178–179
Orientation detection, 185–186, 218
Orthographic neighbourhood size, 72–79
Orthographic representation in letter-by-letter readers, 9, 169, 178–179, 196

Perception deficits, 2, 13–14, 27, 29, 197–198, 214, 216
Perseveration, 157
Phonological priming, 69–71
Picture associations, 215–216
Picture categorisation, 214–215
Picture/object naming, 143, 152, 154, 209, 232–234
    brief exposures, 169
    different perspectives, 214
    exposure time, 179–180
    visual complexity, 27, 29
Picture priming, 220–221
Picture processing, 169, 179–180, 209, 232–234
Priming, case-alternate, 62, 64, 66–67
Priming, identity, 192–194
Priming, phonological, 69–71
Priming, picture, 220–221
Priming in letter-by-letter readers
    covert lexical activation, 57, 58, 66, 67, 81
    homophones, 69–71
    identity priming, 192–194
    picture priming, 220–221
    reading latency, 62–66, 70
    shape-specific knowledge, 58–59
Property judgements, 216
Pseudoletters, 218
Pseudoword recognition, 111–114
Pure alexia *see* Letter-by-letter reading

Reading aloud, 4, 211
Reality decisions, 214, 218
Recovery & rehabilitation of explicit reading, 43, 158
Reverse spelling, 196
Rhyme detection, 147
Right hemisphere
    implicit reading, 61, 142, 150–162
    language processing, 159–161, 228–229
    number processing, 157, 232–234
    picture processing, 154, 232–234
Right hemisphere theories, 4, 12–13, 44–45, 61, 151, 170, 198–199, 204–205, 229–234

Semantic access, 217–218
Semantic categorisation, 10–11, 24, 144–145, 188–189
Semantic priming, 192–194
Sequential processing when reading, 4, 14, 82–83
Severity of letter-by-letter deficit, 42
Shape information & letter/word recognition, 58–60, 84, 144, 228
Spelling, 178–179, 212–213
Split-brains, 60, 157, 159
Spoken word-picture verification, 216
Stimulus degradation
    letter-by-letter readers, 143
    normal readers, 40–41

Tacit reading, 54–56, 61, 80–82, 144–162, 188–195
Tactile naming, 210
Trans-callosal transfer, 3–4
Transcranial magnetic stimulation, 45, 61, 159

Viewing position effect
    children, 127
    factors affecting, 98–99
    letter-by-letter readers, 99–100, 104–105, 113, 122–123
    modelling, 95–98
    normal readers, 95
Visual field asymmetries
    letter recognition, 115, 117
    word length effect, 117
Visual field deficits, 107, 109–110, 113
Visual perception deficits, 2, 13–14, 27, 29, 197–198, 214, 216
Visual processing deficits, 2, 9, 29, 143, 169, 213–218
Visual processing in normal word recognition, 124
Visual span, 181–183, 187–188

White matter, dorsal, 151–152
Word categorisation, 220
Word length effect in letter-by-letter readers, 31, 32, 103–104, 174–175
    imageability, 33, 34, 37, 158
    fixation patterns, 14
    frequency, 33, 34, 37
    presentation time, 149, 177–178
    theories, 127, 129
    variability, 8, 31
    visual field asymmetry, 117
    word vs nonword decisions, 175–177
Word length effect in normal readers, 15, 31–32
    children, 127
    stimulus degradation, 40–41
Word-picture matching, 189–192, 216
Word recognition in letter-by-letter readers
    covert lexical activation, 56–57, 79–80, 81
    parallel vs serial processing, 73–74, 80, 83–84
    shape, 59, 84
Word recognition models, 3, 13, 29, 124, 126
Word superiority effect, 24–25, 55, 147–148
Word vs nonword decisions, 175–177, 220
Writing deficits, 8
Writing to dictation, 178–179